CONTRACTOR LICENSE EXAM PREP:

INTRO:
Embarking on the journey of acquiring a Contractor License is like navigating a ship through uncharted waters. It's thrilling, filled with opportunities, but can also be daunting, and sometimes, the waves of uncertainty and challenge may seem overwhelming. This guide is your compass, here to navigate you through those waters, ensuring that you reach your destination – a successful and fulfilling career as a licensed contractor.

This book is more than just a compilation of information and strategies; it's a beacon of light in times of confusion, a mentor in times of need, and most importantly, a friend who believes in your dreams as much as you do. It understands your aspirations to build, create, and achieve greatness and it validates your fears and uncertainties, standing by you as you face and overcome them.

This guide will not just educate you, but empower you, providing you with the knowledge, skills, and confidence needed to pass the Contractor License exam. It delves deep into the myriad subjects and topics that are essential for the exam, breaking them down into manageable, understandable pieces, making learning not only effective but also enjoyable. It offers real-world examples, practical advice, and interactive elements to help you apply the knowledge in the most effective way, ensuring that your dream is not just possible, but well within your reach.

The journey might be fraught with challenges and moments of doubt, but remember, the path to fulfilling your dreams is worth every effort. It's not just about acquiring a license; it's about building a future, creating opportunities, and achieving the success that you have always aspired to. With every page you turn, you are not just learning; you are evolving, becoming the person you aspire to be, acquiring the skills and knowledge that will be the foundation of your success.

So, delve into this guide with an open heart and a curious mind. Embrace the learning, confront the challenges, and let this book be your companion and guide on your journey to becoming a licensed contractor. Your dreams are valid, your aspirations are worth it, and this guide is your stepping stone to realizing them. Here's to your journey and here's to the success that awaits you!

Gaining a contractor's license is a pivotal step in legitimizing one's construction or contracting business, enhancing credibility and compliance within the construction industry. Licensure demonstrates adherence to industry standards and a commitment to upholding the principles of safety, quality, and professionalism.

Importance of Licensure Acquiring a contractor's license is crucial as it not only validates one's competence and proficiency in construction work but also ensures consumer protection and safety. It facilitates trust among clients, reflecting compliance with local building codes and regulations, and adherence to ethical business practices. It often translates to eligibility for public and private sector projects, expanding business prospects and contributing to professional development.

Contractors are entrusted with significant responsibilities, from overseeing construction activities to ensuring the structural integrity and safety of the projects. Licensure, therefore, acts as a quality assurance mechanism, affirming the contractor's knowledge of building codes, construction principles, and business management, which are pivotal for maintaining industry standards and delivering exemplary services.

Variations in State Requirements It's imperative to understand that contractor licensing requirements exhibit substantial variations across different states. Each state stipulates distinct prerequisites concerning examinations, experience, financial stability, and insurance/bonding requirements. For instance, some states may necessitate contractors to pass both law and trade exams, while others may only require one. The nuances in state-specific regulations underscore the importance of a meticulous approach to understanding and fulfilling the respective licensing criteria.

Let's consider California, where applicants must have at least four years of experience in the construction industry to qualify for the license and must also pass a detailed examination. Conversely, in a state like Florida, contractors must demonstrate financial stability and pass rigorous exams on contractual knowledge and business and finance.

Given the divergent state requirements, it's essential for aspiring contractors to engage in meticulous research or consultation with local licensing boards to discern the precise prerequisites, application processes, and examinations pertinent to their specific jurisdiction.

A real-world illustration of the importance of licensure can be drawn from scenarios where licensed contractors are often preferred by clients for their assurance of quality and compliance, enhancing the contractor's reputation and marketability. For example, a licensed contractor may be contracted to construct a commercial building, where adherence to building codes, safety protocols, and quality standards is non-negotiable. In such instances, licensure is not just a credential; it's a testament to the contractor's capability, reliability, and commitment to industry excellence.

To navigate the labyrinth of varied state requirements effectively, a profound understanding of state-specific laws, building codes, and business practices is paramount. Employing a multifaceted approach, combining self-study, mentorship, practical experience, and continuous learning, can significantly enhance one's preparedness for the licensing exams and subsequent professional endeavors.

This approach is not about merely accruing knowledge; it's about assimilating insights, developing acumen, and cultivating a mindset conducive to professional excellence and ethical conduct in the construction industry. By integrating knowledge, experience, and ethical considerations, aspiring contractors can transcend conventional learning paradigms, positioning themselves as vanguards of industry standards and advocates of consumer interests.

Navigating eligibility criteria is like meticulously assembling the pieces of a puzzle. Each component, from age to experience to education, plays a crucial role in framing the overall picture, and varying state requirements add another layer of complexity to this intricate process.

Age, Experience, and Educational Qualifications: Most states mandate applicants to be at least 18 years of age and hold a high school diploma or equivalent. The significance of experience is paramount; typically, a range of 2 to 4 years of relevant, practical experience in the construction industry is a prerequisite. This experience is a testament to an individual's hands-on exposure to construction work, knowledge of construction principles, and understanding of safety protocols. Educational qualifications play a pivotal role too; some states offer substitutions for experience with a degree in construction or a related field. For instance, holding a bachelor's degree in construction management might alleviate some of the experience requirements, showcasing an individual's comprehensive understanding of construction theory, project management, and business practices.

Delving into state-specific prerequisites reveals a spectrum of distinct requirements and processes. In states like Virginia, applicants must substantiate their financial stability and showcase proof of insurance and bonding. Additionally, certain states have reciprocity agreements allowing licensed contractors from one state to obtain a license in another without retesting, provided the states have analogous licensing requirements. A meticulous review of state licensing boards' stipulations is pivotal for accurate, up-to-date insights on eligibility criteria, application processes, and examination requirements.

For instance, Oregon has stringent prerequisites necessitating contractors to complete pre-license training and pass a state exam. Conversely, in Georgia, contractors need to exhibit their financial solvency and pass exams focusing on business law and construction skills.

In real-world terms, these eligibility criteria act as the gatekeepers of the profession. A young professional, brimming with knowledge from a degree in construction engineering, must augment this learning with practical, on-site experience, melding theoretical insights with hands-on skills. Balancing the scales between education and experience can streamline the journey towards licensure, facilitating a seamless transition from learning to practicing, ensuring the prospective contractor is well-versed in both the academic and pragmatic aspects of construction work.

It's crucial to regard these eligibility criteria not as hurdles but as building blocks, each shaping a prospective contractor's knowledge base, skillset, and professional ethos. A multifaceted approach, intertwining education, experience, and continuous learning, enables aspiring contractors to cultivate a profound understanding of construction principles, business acumen, and ethical considerations, fostering professional growth and contributing to the elevation of industry standards. The amalgamation of these elements doesn't just prepare individuals for licensure but molds them into competent, ethical, and knowledgeable professionals, capable of navigating the multifarious landscape of the construction industry with adeptness and integrity.

Embarking on the application process necessitates precision, attentiveness, and a comprehensive understanding of each step, ensuring accuracy and compliance with all respective requirements.

Submitting Applications:
Applicants must thoroughly complete the necessary forms, generally provided by the state's licensing board or its equivalent. Attention to detail is crucial, as inaccuracies or omissions can lead to delays or rejection. Some states offer online submission portals, streamlining the application process and facilitating prompt communication between the applicant and the licensing board.

Required Documentation:
Documentation serves as the backbone of the application process. Generally, proof of age, identification, educational qualifications, and relevant experience must be meticulously compiled. Verification of experience, often through references or employer testimonials, is a pivotal component, substantiating the applicant's hands-on experience in the field. Financial statements and proof of insurance may also be requisite, depending on state-specific regulations. It's imperative to cross-verify each document, ensuring its validity and relevance to the application.

Fees are an integral aspect of the application process. These can encompass application fees, examination fees, and licensing fees. It's paramount to review the fee structure meticulously, as it varies significantly across states. For instance, some states may charge a nominal fee for the application but impose a substantial fee for the license issuance. Fee waivers or reductions might be available in certain circumstances, and it's worthwhile to explore such provisions to alleviate financial burden.

Imagine a scenario where a contractor, adept in construction principles and equipped with extensive experience, is poised to submit an application. This individual must synchronize every piece of information, from educational credentials to professional references, presenting a cohesive, accurate depiction of their professional journey. The submission of accurate, verifiable documentation is not just a procedural necessity but a reflection of the applicant's professionalism and commitment to ethical practices.

The application process is akin to laying the foundation of a structure. It demands meticulousness, integrity, and adherence to protocols, shaping the trajectory of an aspiring contractor's professional journey. This step is not merely a gateway to licensure but a manifestation of an individual's dedication to their craft, adherence to ethical standards, and commitment to continuous learning and professional development. The meticulous orchestration of each element of the application process reinforces the sanctity of the profession, fostering a culture of competence, accountability, and excellence in the construction industry.

Business and Law:

When embarking on a contracting venture, choosing the right business structure is pivotal. It's like laying the foundation for a structure, where a solid base can foster growth, stability, and sustainability.

Sole Proprietorship

The simplest form is a sole proprietorship. Here, the business and owner are legally the same entity. While it's easy to set up, the owner assumes all the liability. Any legal actions taken against the business target the owner's personal assets. The income earned is also considered the owner's personal income and is taxed accordingly, often leading to higher tax brackets.

Partnership

Then there is a partnership, where two or more individuals share ownership. Each partner contributes to all aspects of the business, including money, property, labor, or skill. Each partner shares the profits and losses of the business, and like the sole proprietorship, personal assets can be pursued in legal actions. Partnerships must file an annual information return to report income, deductions, gains, and losses, but the business itself does not pay income tax. Instead, the profits and losses are passed through to the partners.

Corporation

A corporation is a more complex structure, considered a separate entity from its owners, providing liability protection but involving rigorous record-keeping, operational processes, and reporting. Corporations are subject to double taxation, once on corporate profits and again when distributed to shareholders. However, corporations have advantages in raising capital and continuity.

S Corporation

An S Corporation offers avoidance of double taxation, with income, deductions, and credits flowing through to shareholders. It provides liability protection but is subject to numerous restrictions, including the number and type of allowable shareholders.

Limited Liability Company (LLC)

Limited Liability Company (LLC) combines the liability protection of a corporation with the tax benefits and operational flexibility of a partnership. Owners, called members, report profits and losses on their personal income tax returns, avoiding double taxation.

Choosing the Right Structure

The choice of business structure significantly influences daily operations, the ability to raise capital, the paperwork and operational formalities required, personal liability, and how much one pays in taxes. A well-considered decision can protect personal assets, offer a favorable tax scenario, ease the fundraising process, and ultimately, pave the way for success.

Understanding the nuances of each structure, aligning it with the business goals, and adjusting as the business evolves are crucial steps to ensure that the chosen business structure serves the strategic objectives and operational needs of the contracting business, and is aligned with long-term vision and growth.

Remember, the information provided does not serve as legal or tax advice, and individuals should consult with qualified professionals or legal counsel to determine the most suitable business structure for their specific needs and circumstances.

In a contracting business, operations management is like the central nervous system, efficiently integrating and coordinating various functions to achieve optimal productivity and profitability. It involves meticulous planning, organizing, and supervising the processes of production and manufacturing.

Supply Chain Management

In contracting, supply chain management is crucial. It's about managing the flow of goods and services, ensuring that everything from the procurement of raw materials to the delivery of the end product is seamless. An efficient supply chain is characterized by timely deliveries and an absence of excess inventory, and it significantly impacts the cost structure and profitability of the contracting business. It necessitates robust vendor relationships, effective negotiation skills, and rigorous contract management to mitigate risks related to cost overruns, delays, and quality deficiencies.

Logistics

Logistics is the heartbeat of a contracting business, focusing on the coordination and movement of resources. A well-orchestrated logistics plan is paramount in avoiding delays and ensuring the timely availability of manpower, equipment, and materials, directly influencing project timelines and profitability. Implementing technology solutions for scheduling, routing, and tracking can significantly enhance logistical efficiency, reducing downtime and avoiding bottlenecks.

Quality Control

Quality control in the contracting field is non-negotiable. It's the cornerstone of delivering value, sustaining business reputation, and avoiding costly rework and liability issues. It involves stringent standards, systematic inspection, testing, and documentation to ensure that the workmanship and materials conform to the predetermined specifications and standards. Employing a proactive approach, identifying potential issues before they escalate, and instigating corrective actions are paramount in maintaining quality standards.

Optimizing Operational Efficiency

Optimizing operational efficiency is about refining these components to function like a well-oiled machine. It involves leveraging technology, implementing lean practices to minimize waste, enhancing productivity through workforce management, and fostering a culture of continuous improvement. Lean construction techniques such as Just-In-Time delivery, Value Stream Mapping, and 5S methodology can be instrumental in eliminating inefficiencies and maximizing value.

Imagine orchestrating the construction of a multi-story building. A minor delay in material delivery can halt the entire operation, leading to cascading delays and cost overruns. By employing precise logistics and supply chain strategies, such as advanced scheduling and real-time tracking, you can ensure that materials arrive just when needed, avoiding delays and excess inventory costs. Implementing rigorous quality control practices ensures the longevity and integrity of the building, reducing the likelihood of future structural issues and associated liability risks.

Meticulous attention to operations management principles is pivotal in navigating the complex and dynamic environment of the contracting business, enabling the delivery of high-quality projects on time and within budget, while maximizing profitability and sustaining business growth.

Contract law in construction is the bedrock upon which projects are built, involving agreements that are legally binding and which delineate the obligations and rights of the involved parties. For contractors, understanding these principles is crucial to safeguarding interests and mitigating risks.

Fundamental Principles

1. **Offer and Acceptance:** Every contract begins with an offer by one party and acceptance by another. This mutual consent is essential, and any discrepancies can lead to a void contract.
2. **Consideration:** It represents the value exchanged between parties, typically in the form of services, money, or goods, establishing the contract's enforceability.
3. **Legal Purpose:** Contracts with unlawful objectives are invalid. For construction contracts, adhering to zoning laws, building codes, and industry standards is imperative.
4. **Certainty and Possibility of Performance:** Terms must be clear, and performance must be feasible. Ambiguous terms or impossibility of performance can invalidate the contract.

Legally Sound, Equitable, and Enforceable Contracts

1. **Detailed Scope of Work:** Clearly defining the work scope, including tasks, deliverables, and milestones, eliminates ambiguities and establishes accountability.
2. **Fair and Clear Terms:** All terms, including payment schedules, completion dates, and penalties, should be transparent, reasonable, and mutually agreed upon.
3. **Risk Allocation:** Risks like unforeseen site conditions or regulatory changes should be equitably allocated, with clear mechanisms for addressing them.
4. **Dispute Resolution:** Specifying methods for resolving disputes, such as arbitration or litigation, provides a predefined path for addressing disagreements.
5. **Compliance with Laws:** Ensuring adherence to all relevant laws and regulations, including licensing and permitting requirements, validates the contract's legality.
6. **Consultation with Legal Counsel:** Having contracts reviewed by legal experts ensures their soundness, compliance with laws, and enforceability.

Consider a contract for constructing a commercial building. Having clear and comprehensive contracts detailing every aspect, from material specifications to payment terms, minimizes disputes and ensures smooth project execution. If disputes arise, predetermined resolution mechanisms facilitate quicker, more amicable resolutions, avoiding prolonged legal battles and maintaining healthy business relationships.

Contract law's intricate weave requires contractors to be vigilant, knowledgeable, and proactive, creating contracts that are clear, fair, and enforceable, thus laying a solid foundation for successful project completion and long-term business sustainability. Balancing legal acumen with operational insight is key in navigating the contractual landscapes of the construction industry.

In the construction industry, knowledge of employment and labor laws is non-negotiable. This knowledge ensures a harmonious workplace, avoids legal disputes, and maintains organizational integrity.

Essential Labor Laws

1. **Fair Labor Standards Act (FLSA):** Governs minimum wage, overtime pay, and working hours, ensuring workers receive fair compensation.
2. **Occupational Safety and Health Act (OSHA):** Stipulates safety and health standards to protect workers from potential hazards on job sites.
3. **Family and Medical Leave Act (FMLA):** Allows eligible employees to take unpaid, job-protected leave for specified family and medical reasons.
4. **National Labor Relations Act (NLRA):** Protects the rights of employees to act together to address conditions at work, with or without a union.

Staying Compliant

1. **Regular Training and Updates:** Regularly train employees and management on their rights and responsibilities and stay abreast of any changes in labor laws.
2. **Implement Safety Protocols:** Adopt and rigorously enforce safety standards to minimize workplace accidents and OSHA violations.
3. **Maintain Proper Recordkeeping:** Proper documentation of employee hours, wages, and other required details is crucial for compliance with FLSA and other laws.
4. **Transparent Employment Policies:** Clear, accessible, and well-communicated policies ensure employees understand their rights and responsibilities.
5. **Appropriate Worker Classification:** Correctly classifying workers as employees or independent contractors avoids misclassification penalties.

Consider a contractor employing a team for a residential building project. Implementing regular safety trainings and maintaining meticulous work records not only ensures compliance with OSHA and FLSA but also fosters a positive work environment and minimizes the risk of legal repercussions. Clear communication of employee rights and employer obligations reinforces trust and transparency within the organization.

- Establishing a dedicated compliance team or consultant can significantly mitigate legal risks by ensuring continuous adherence to employment and labor laws.
- Developing robust internal communication channels can facilitate better understanding and adherence to employment policies and labor laws among employees.

Delving deep into labor laws, embracing proactive compliance strategies, and fostering an environment of transparency and safety can help contractors navigate the intricate landscape of employment and labor laws effectively, building a solid, legally compliant operational framework.

Grasping fundamental accounting principles is vital for contractors to maintain fiscal health and compliance.

Essential Accounting Principles

1. **Accrual Basis Accounting:** Recognizing income and expenses when they are earned or incurred, rather than received or paid.

2. **Matching Principle:** Aligning expenses with the revenue they generate within the same accounting period.
3. **Revenue Recognition Principle:** Recording revenue when earned, not necessarily when received.
4. **Cost Principle:** Recording assets at their cost price and not their market value.

Managing Taxation

1. **Maintain Accurate Records:** Keeping meticulous records of income, expenses, and asset acquisitions is fundamental for accurate tax filings.
2. **Understand Tax Deductions:** Knowledge of allowable deductions, such as business expenses and depreciation, can optimize tax liability.
3. **Regular Tax Planning:** Engaging in periodic tax planning can identify strategies to minimize tax liability, such as leveraging tax credits and deductions.
4. **Hire a Certified Accountant or Tax Professional:** Engaging professionals can ensure proper compliance with tax laws and optimization of tax positions.

For example, a contractor purchasing construction equipment should record the transaction at the cost price and appropriately depreciate the asset over its useful life, utilizing the matching principle to allocate the expense against the revenue generated by the project.

- Employing accounting software can facilitate accurate and efficient record-keeping and financial reporting, enabling contractors to monitor financial health and make informed business decisions.
- Leveraging tax-advantaged investment vehicles and retirement plans can offer significant tax savings, aligning financial strategies with long-term business goals.

Contractors who master the intricacies of accounting principles and taxation not only ensure legal compliance and financial stability but also strategically position their businesses for sustained growth and success. By maintaining precise financial records, understanding applicable tax laws, and implementing strategic financial planning, contractors can optimize their financial operations and navigate the complexities of the financial landscape in the construction industry efficiently.

Crucial Aspects of Risk Management

Risk management in contracting is pivotal; it involves systematically identifying, assessing, and mitigating risks to avoid fiscal and operational disruption.

1. **Risk Identification:** Contractors should continuously identify potential risks, such as construction defects, delays, or cost overruns, inherent in each project phase.
2. **Risk Assessment:** Once identified, risks should be evaluated concerning their likelihood and potential impact, enabling prioritization and effective management.
3. **Risk Mitigation:** Implementing proactive strategies, such as contract clauses limiting liability and comprehensive safety programs, can mitigate identified risks.

Importance of Insurance in Construction

Insurance is paramount in managing the multifaceted risks inherent in the construction industry, shielding contractors from substantial financial losses.

1. **General Liability Insurance:** Essential for covering property damage and bodily injuries arising from construction activities.
2. **Workers' Compensation Insurance:** Mandated by law, it covers medical expenses and wage replacement for employees injured on the job.
3. **Professional Liability Insurance:** Protects against claims arising from errors, omissions, or negligence in the provision of professional services.
4. **Builder's Risk Insurance:** Covers property damage to ongoing construction projects due to perils like fire, wind, theft, and vandalism.
5. **Commercial Auto Insurance:** Essential for vehicles operated by the business, covering liabilities arising from accidents.
6. **Surety Bonds:** Not insurance per se, but they assure project owners of the contractor's capability to fulfill contractual obligations.

Real-world Applications and Deep Insight

For instance, maintaining stringent safety protocols can significantly diminish the risk of worksite accidents, thus reducing liability exposures and potential Workers' Compensation claims. Similarly, meticulously drafting contracts can alleviate disputes related to construction delays or defects, mitigating potential legal ramifications.

By recognizing the pivotal role of risk management and insurance, contractors can preemptively address potential adversities, ensuring project continuity and financial stability. Ensuring apt insurance coverage not only shields against unforeseen financial strains but also fosters credibility and trust amongst clients and partners in the construction industry.

Developing an in-depth, comprehensive risk management plan, routinely reassessing risks, and adapting insurance coverage accordingly, ensures contractors are continually safeguarded against the evolving risk landscape in construction. Leveraging industry-specific risk management tools and technologies can offer revolutionary insights, enabling contractors to predict, prioritize, and prevent potential risks proactively.

Leveraging technology like predictive analytics can provide advanced risk foresight, allowing for proactive risk mitigation strategies, such as optimized scheduling to avoid weather-related delays or advanced safety training to address identified accident-prone activities. This proactive approach in leveraging technology for risk management can significantly minimize the occurrence and impact of construction-related risks.

Contractors who embrace robust risk management and maintain adequate insurance fortify their business against unforeseen impediments, ensuring the seamless progression and completion of construction projects. By understanding and implementing advanced and comprehensive risk management strategies and securing appropriate insurance coverages, contractors pave the way for operational excellence, financial stability, and long-term success in the highly competitive and dynamic construction industry.

Best Practices for Hiring

When hiring in the contracting business, it is pivotal to seek candidates with the aptitude, experience, and work ethic that align with the company's values and objectives.

1. **Clear Job Descriptions:** Clearly outline roles, responsibilities, qualifications, and expectations to attract the right candidates.
2. **Rigorous Screening:** Employ rigorous screening processes, including background checks and reference validations, to ensure the reliability and integrity of potential hires.
3. **Skill Assessment:** Assess technical and soft skills to ensure candidates possess the requisite expertise and interpersonal acumen needed for the role.

Training and Employee Development

Optimal training and employee development are instrumental in nurturing a skilled, knowledgeable, and motivated workforce.

1. **Comprehensive Training Programs:** Develop and implement comprehensive training programs to equip employees with the necessary skills and knowledge, emphasizing safety and quality standards.
2. **Continuous Learning Opportunities:** Provide access to ongoing learning opportunities, such as workshops, seminars, and courses, to foster employee growth and adaptability.
3. **Performance Evaluations and Feedback:** Regularly assess employee performance and provide constructive feedback to facilitate continuous improvement and career development.

Contribution to Business Growth and Success

Effective human resources management is indispensable for the growth and success of a contracting business, contributing to enhanced productivity, operational excellence, and employee retention.

1. **Enhanced Productivity:** A well-hired and adequately trained workforce is more productive and efficient, contributing to project success and client satisfaction.
2. **Operational Excellence:** Effective HR practices cultivate a positive and conducive work environment, promoting operational excellence and adherence to quality and safety standards.
3. **Employee Retention:** A supportive and growth-oriented environment fosters employee satisfaction and loyalty, reducing turnover and associated costs.

Consider a scenario where a contracting company invests in advanced technical training for its employees. This investment not only enhances the skillset of the workforce but also elevates the company's service quality and market competitiveness, ultimately contributing to increased client acquisition and retention.

Empowering employees through structured development programs, aligned with cutting-edge industry practices and technologies, can propel a contracting business to unprecedented heights. By fostering a culture of continuous learning and improvement, contractors can stay abreast of industry advancements, ensuring sustained growth and competitiveness in the evolving market landscape.

Developing a mentorship program where experienced workers guide newer hires can facilitate rapid skill acquisition and acclimatization to the company culture. Leveraging technological solutions for training, such as virtual reality for safety training, can provide immersive learning experiences, enhancing knowledge retention and application.

Effective hiring, meticulous training, and continuous employee development are the linchpins of successful human resources management in contracting. By adopting best practices in these areas, contractors can cultivate a competent, motivated, and cohesive workforce, laying the foundation for organizational growth, client satisfaction, and industry leadership.

Core Wage Laws
In the construction industry, several critical wage laws are designed to protect the rights and incomes of employees.
1. **Fair Labor Standards Act (FLSA):** It mandates minimum wage, overtime pay eligibility, recordkeeping, and child labor standards.
2. **Davis-Bacon Act:** It requires contractors to pay prevailing wages for federally-funded construction projects.
3. **State Labor Laws:** They can vary but typically dictate minimum wage, overtime, and break periods, which may be more stringent than federal laws.

Employee Rights
Employees in the construction sector have a multitude of rights focusing on fair compensation, safe working conditions, and freedom from discrimination.
1. **Right to Fair Compensation:** Employees are entitled to receive at least the minimum wage and are entitled to overtime pay for hours worked beyond the standard workweek.
2. **Right to a Safe Workplace:** Employees have the right to work in environments where risks are properly managed and mitigated according to OSHA standards.
3. **Right to Equal Opportunity:** Employees have the right to fair and equal treatment, free from discrimination or harassment based on race, color, religion, sex, or national origin under Title VII of the Civil Rights Act.

Compliance with Wage Laws and Employee Rights
Contractors can ensure compliance with wage laws and respect employee rights through several best practices:
1. **Knowledge and Understanding:** Regularly update knowledge of federal and state wage laws to ensure compliance.
2. **Accurate Recordkeeping:** Maintain precise records of hours worked, wages paid, and other pertinent employment details to verify lawful compensation.
3. **Transparent Communication:** Clearly communicate wage rates, payment schedules, and employment terms to employees and address any queries or concerns promptly.
4. **Regular Training:** Train HR and management teams on legal requirements and updates relating to wage laws and employee rights.
5. **Compliance Audits:** Periodically conduct internal audits to verify adherence to all applicable wage laws and rectify any discrepancies identified.

In real-world settings, contractors utilizing accurate time-tracking software and maintaining clear, open lines of communication regarding compensation and expectations can prevent misunderstandings and disputes, fostering a harmonious and compliant workplace environment.

The integration of advanced technological solutions, like blockchain, for payroll can revolutionize compliance with wage laws. Blockchain can offer immutable, transparent, and accurate payroll processing, ensuring every employee is compensated correctly, reducing disputes and enhancing trust within the organization.

Implementing proactive compliance strategies, such as seeking legal counsel for contract reviews and wage compliance evaluations, can help contractors avoid potential legal pitfalls. By fostering an environment where employee rights are paramount, contractors can not only ensure legal compliance but also improve employee morale and productivity, which in turn, can lead to improved project outcomes and business success.

Construction Practices and Techniques:

Structural elements are the backbone of buildings, ensuring stability, endurance, and resilience. The foundational elements include columns, beams, and slabs, intricately connected, acting in unison to bear and distribute loads efficiently, preventing structural failure.

1. Columns:

Columns are vertical structural members, crucial for bearing the building's load, transferring it to the foundations. Typically, reinforced concrete, steel, or timber is employed, chosen based on load requirements, economic feasibility, and architectural considerations. Correct column placement is pivotal, demanding meticulous calculations and precise execution, optimizing structural integrity.

2. Beams:

Beams are horizontal members, predominantly bearing live and dead loads, subsequently transferring them to columns. Material selection, spanning from steel to reinforced concrete, is contingent upon load specifications, environmental conditions, and design requisites. Incorporating adequate beam size and reinforcement is paramount, mitigating risks of structural deformations and failures.

3. Slabs:

Slabs form the floors and ceilings, spreading loads over an area, directing them to the beams. Reinforced concrete is prevalently utilized, prized for its versatility and strength. Slabs' thickness and reinforcement are meticulously calculated, considering load intensities and spans, ensuring durability and safety.

4. Foundations:

Foundations anchor structures to the ground, absorbing and distributing loads to the earth. Deep or shallow foundations are adopted based on soil characteristics, load magnitude, and site conditions. Proper foundation design and construction are imperative, averting disproportionate settlements and structural instability.

5. Trusses and Frames:

Trusses and frames, composed of triangles and polygons respectively, are employed to span large distances, ideal for roofs and bridges. Material choices range from timber to steel, influenced by span length, load requirements, and aesthetic preferences. The configuration and connection of components require precise engineering, ensuring load distribution and structural robustness.

6. Walls and Partitions:

Walls act as vertical enclosures, supporting roofs and floors, while partitions subdivide spaces. Constructed from bricks, blocks, or framed structures, their design considers load-bearing capacities, fire resistance, and acoustic properties. Proper integration of these elements is crucial for structural coherence and functionality.

Integration Methods:

Integrating these elements necessitates a cohesive and holistic approach, involving meticulous detailing, accurate positioning, and robust connections. Employing advanced software tools for structural analysis and design is commonplace, enabling precise calculations and optimal material utilization.

These elements, interconnected and harmoniously integrated, collectively uphold the building, resisting various forces, from gravitational to lateral loads, securing longevity and stability. Mastery over the knowledge and application of these elements is indispensable for aspiring contractors, aligning with codes and standards, fostering structural excellence and safety. Remember, adhering to local building codes and integrating advancements in structural engineering will not only ensure the longevity and safety of the structure but will also facilitate smoother project execution and approval processes, a crucial aspect for any contractor aiming for excellence in the field.

The building envelope serves as the intermediary between the interior and the external environment, playing a pivotal role in sheltering the interior from weather, controlling the internal climate, and providing structural support. It predominantly consists of the roof, walls, windows, doors, and foundation.

1. Roof:

The roof, the uppermost shield, safeguards against environmental elements, such as precipitation and sunlight. Employing materials like asphalt shingles, metal, or tiles, it's crucial to integrate proper insulation and waterproofing to avert water ingress and energy loss.

2. Walls:

Walls are the vertical components, encircling and defining the structure's perimeter. Constructed from materials like bricks, concrete, or timber, they necessitate adequate insulation and vapor barriers to combat heat loss and moisture accumulation.

3. Windows and Doors:

Windows and doors facilitate ingress, egress, and daylight penetration. Their design and installation must counteract energy loss, utilizing energy-efficient glazing and sealing solutions, optimizing thermal performance and comfort levels.

4. Foundation:

The foundation, the structure's base, must be impermeable and robust, preventing ground water infiltration and supporting structural loads. Incorporating damp-proofing and insulation is imperative, enhancing thermal efficiency and structural longevity.

5. Building Wrap and Insulation:

A building wrap acts as a protective layer, mitigating air and water infiltration, while allowing vapor transference. Employing adequate insulation in walls, roofs, and floors is paramount, minimizing thermal bridges and energy consumption.

Contractors' Role in Building Envelope Effectiveness:

Contractors are responsible for ensuring the envelope's effectiveness in protection, insulation, and energy efficiency. Proper material selection, meticulous installation, and compliance with energy codes and standards are critical. Employing techniques such as thermal bridging analysis and air leakage testing can verify the envelope's performance, pinpointing areas of improvement. Contractors must coordinate with architects and engineers, adopting integrated design approaches and leveraging advancements in building science, achieving envelopes that are resilient, energy-efficient, and compliant with sustainability goals.

Energy Efficiency and Sustainability:

Optimizing the building envelope is crucial for energy conservation and sustainability. An effective envelope reduces HVAC loads, curtails energy expenditure, and mitigates carbon emissions. Incorporating renewable energy solutions, such as solar panels on roofs, can further enhance energy efficiency and sustainability.

The building envelope's comprehensive and well-executed design and construction are indispensable for any contractor aiming for high-quality and sustainable construction. It necessitates a profound understanding of materials, construction techniques, and building physics, ensuring the envelope performs optimally, safeguarding the building and its occupants while contributing to environmental sustainability.

In essence, the correct implementation of a building envelope is pivotal in the creation of structures that are both resilient and energy-efficient, aligning with the contemporary emphasis on sustainability and energy conservation in the construction industry.

Site work is a fundamental phase in construction, laying the groundwork for a project and involving several critical steps and considerations before actual building commences. It sets the tone for the subsequent construction phases and demands meticulous attention to ensure the stability and longevity of the structure.

1. Site Assessment and Survey:

An initial site assessment is crucial, identifying potential environmental and logistical challenges. It incorporates a topographical survey to understand the land's contours, elevations, and existing features, impacting the design and placement of the structure.

2. Soil Testing:

Soil testing is indispensable, determining the soil's type, composition, density, and bearing capacity. The soil's characteristics influence the foundation type, whether it's a shallow or deep foundation, and necessitate suitable soil stabilization and improvement techniques if the soil is found to be inadequate.

3. Site Clearing:

This involves the removal of vegetation, debris, and existing structures, preparing the land for construction. Proper disposal or recycling of cleared materials is crucial to adhere to environmental regulations.

4. Grading and Excavation:

Grading ensures a level base or specifies slope for construction and manages water runoff. Excavation is performed to create space for foundations, with attention to soil stability and measures to prevent collapse of excavated areas.

5. Utilities and Services Installation:

Before construction, it's imperative to layout and install essential utilities like water, electricity, and sewer lines. Adequate planning prevents conflicts and ensures that services are accessible for connection.

6. Foundation Work:

Based on soil conditions and structural requirements, foundation work is carried out to transmit loads to the ground securely. Ensuring proper footing and foundational support is critical to avoid structural failures.

Impact of Site Conditions, Soil Type, and Topography:

Site conditions, soil type, and topography significantly impact construction methodologies, costs, and timelines. The presence of rocky or clayey soil, high water table, or steep slopes may necessitate specialized construction techniques and additional measures for stabilization and drainage. Adverse conditions may lead to increased construction time and costs due to the need for more extensive excavation, soil improvement, and foundation works.

Practical Considerations:

For instance, a site with a steep slope will demand extensive grading and retaining structures to create a stable and level base, impacting the project's cost and timeline. Conversely, a flat site with sandy soil may allow for rapid progress but might require soil compaction and pile foundations to ensure stability.

Meticulous attention to site work is indispensable for the successful execution of a construction project. A profound understanding of the site conditions, topographical features, and soil characteristics enables the formulation of effective construction strategies, mitigating risks, and ensuring the structural integrity and viability of the project. Contractors must leverage precise assessments, innovative solutions, and effective coordination to navigate the complexities of site work, laying a solid foundation for the subsequent phases of construction.

Concrete:

Concrete is a composite material composed of fine and coarse aggregate bonded together with a fluid cement that hardens over time. It is incredibly durable and strong, especially against compression. In construction, it's utilized for structural elements like foundations, walls, and slabs. It's crucial to properly mix and cure concrete to ensure its strength and longevity.

Application: It's used predominantly for its strength, moldability, and ability to encase steel reinforcements, providing tensile strength.

Handling: Concrete must be kept moist while curing to reach its full strength potential, usually for at least 28 days, and should be properly mixed, taking into account the water-cement ratio.

Masonry:

Masonry involves constructing structures by laying individual masonry units like bricks, stones, or concrete blocks. It's celebrated for its fire resistance, durability, and aesthetic versatility.

Application: Used for building walls, partitions, and retaining structures, offering aesthetic flexibility and structural robustness.

Handling: Proper bonding, accurate leveling, and adequate mortar mixing are essential to ensure structural integrity and longevity.

Metals:

Metals such as steel and aluminum are crucial in construction due to their tensile strength, ductility, and durability.

Application: Steel is predominantly used for structural frameworks, reinforcements, and foundations, while aluminum is used for window frames, doors, and panels due to its lightweight and corrosion resistance.

Handling: Metals should be adequately protected against corrosion, and connections should be robust, utilizing welding, bolting, or riveting.

Wood:

Wood is an organic material that is structurally strong and aesthetically pleasing, offering thermal and acoustic insulation.

Application: Employed for framing, flooring, and roofing, providing structural support and aesthetic appeal.

Handling: Wood must be adequately treated to protect against decay, insects, and fire, and should be appropriately seasoned to prevent warping and shrinkage.

Plastics:

Plastics are synthetic polymers that are lightweight, corrosion-resistant, and versatile.

Application: Utilized for piping, cladding, and insulation due to their versatility and resistance to various environmental conditions.

Handling: Plastics should be selected based on their specific properties, ensuring suitability for the intended application, and should be properly installed to avoid deformations and failures.

Impact on Quality and Durability:

The proper selection and use of materials significantly impact the quality, durability, and performance of the construction project. Using materials with inadequate strength or durability can lead to structural failures, while improper installation or handling can reduce the effective lifespan of the materials and the overall structure. Conversely, choosing high-quality materials and employing proper construction techniques enhance the structure's resilience, longevity, and functionality.

Consider a high-rise building; using high-strength concrete for the foundation and structural elements, corrosion-resistant steel for reinforcements, and high-quality bricks for façades, combined with meticulous construction practices, will ensure the building's stability, durability, and aesthetic appeal.

To sum up, every material has unique properties that determine its suitability for specific applications in construction. A deep understanding of these properties, coupled with correct handling and installation, is crucial for constructing structures that meet the desired standards of quality and durability. Balancing material selection, construction techniques, and cost considerations while adhering to building codes and standards is the hallmark of effective construction practices.

Mechanical Systems:

Mechanical systems like Heating, Ventilation, and Air Conditioning (HVAC) regulate the internal environment of a building.

Considerations: Efficient design and proper sizing are critical for energy efficiency and comfort. Contractors need to consider load calculations, energy consumption, and system controls.

Best Practices: Regular maintenance, proper installation, and ensuring optimal airflow are essential. The use of energy-efficient appliances and the integration of renewable energy sources can enhance efficiency.

Compliance and Safety: Adherence to ASHRAE standards and local building codes is crucial to ensure safety and performance.

Electrical Systems:

Electrical systems provide power to buildings, enabling the operation of lights, appliances, and other equipment.

Considerations: Adequate capacity, load balancing, and future expansion possibilities should be taken into account during the design and installation phase.

Best Practices: Regular inspection of wiring and components, use of high-quality materials, and ensuring proper grounding are paramount. Circuit breakers should be appropriately labeled, and safety devices like GFCIs should be installed where needed.

Compliance and Safety: Compliance with the National Electrical Code (NEC) and local regulations is non-negotiable to prevent fire risks and electrical failures.

Plumbing Systems:

Plumbing systems supply water and remove waste, contributing to the building's hygiene and comfort.

Considerations: Proper pipe sizing, material selection, and slope are critical to prevent blockages and ensure adequate water pressure.

Best Practices: Leak prevention, insulation of pipes, and regular maintenance can prevent water damage and enhance efficiency. Water-saving fixtures and appliances should be prioritized to conserve water.

Compliance and Safety: Adherence to the Uniform Plumbing Code (UPC) and local codes is crucial to ensure safe and sanitary plumbing installations.

Fire Protection Systems:

Fire protection systems like sprinklers and alarms are crucial for safeguarding buildings and occupants from fire hazards.

Considerations: Proper placement of detectors, adequate water supply for sprinklers, and regular testing are vital for effective fire protection.

Best Practices: Installing fire-resistant materials, maintaining clear escape routes, and educating occupants on fire safety are also crucial. Fire protection systems should be integrated with alarm and notification systems.

Compliance and Safety: Compliance with the National Fire Protection Association (NFPA) standards and local fire codes is mandatory to ensure the safety of the building and its occupants.

Ensuring Compliance, Efficiency, and Safety:

Contractors should rigorously follow design specifications and installation procedures, conduct regular inspections, and work closely with inspectors to ensure all systems comply with applicable codes and standards. Energy-efficient designs, regular maintenance, and the use of high-quality materials contribute to system efficiency and safety. Employing licensed and experienced subcontractors and workers for specialized tasks is also key to avoiding errors and ensuring the highest quality workmanship.

In the construction of a commercial building, the meticulous integration of mechanical, electrical, plumbing, and fire protection systems is crucial. A flaw in the electrical system can lead to catastrophic results, while an inefficient HVAC system can result in high energy bills and discomfort for the occupants. A well-designed plumbing system ensures sanitation, and a robust fire protection system can be the difference between life and death in emergency situations. Each of these systems plays a pivotal role in the functionality, safety, and comfort of buildings. A well-rounded knowledge of the principles, best practices, and compliance requirements associated with each is indispensable for contractors aiming for excellence in building construction.

Paving Principles and Techniques:
For paving, the substratum preparation is paramount, requiring meticulous grading, compaction, and drainage planning to prevent water accumulation and ensure longevity. The choice of paving materials like asphalt, concrete, or pavers is determined by traffic load, aesthetic preference, and budget. Proper installation techniques such as achieving optimal compaction and joint placement in concrete are pivotal to avoid premature deterioration.

Landscaping Principles and Techniques:
In landscaping, soil analysis is the foundation for selecting suitable plant species and determining irrigation needs. The design should incorporate a mix of plant types and heights to create visual interest and biodiversity. Consideration of sunlight, wind direction, and microclimate is crucial for plant health and placement of features like patios and pergolas.

Balancing Aesthetics, Functionality, and Sustainability:
Contractors leverage design principles like balance, contrast, and unity to achieve aesthetically pleasing paving and landscaping. Functional considerations include user accessibility, traffic flow, and maintenance ease. Sustainable practices encompass the use of permeable paving materials to reduce runoff, native and drought-resistant plants to conserve water, and eco-friendly landscape materials to minimize environmental impact.

Practical Application and Real-World Example:
Consider a public park project. Contractors need to ensure that paths and walkways are durable and accessible, possibly opting for permeable pavers to manage stormwater runoff effectively. Landscaping should be attractive, with plantings that provide shade, habitat, and seasonal interest, using native and adaptive species to reduce water and maintenance needs.

Expert Insight and Uncommon Strategies:
An advanced strategy is the integration of green infrastructure, like rain gardens and bioswales, to manage stormwater on-site, enhancing sustainability. The implementation of advanced irrigation systems like drip irrigation and smart controllers can optimize water use. For paving,

employing innovative materials like high-performance concrete or permeable asphalt can enhance durability and sustainability.

Contractor Insight and Professional Practices:

Professionals understand that the synergy of aesthetics, functionality, and sustainability in paving and landscaping is essential for project success. They continually upgrade their knowledge on new materials, technologies, and best practices to deliver optimal outcomes. This includes keeping abreast of evolving sustainability standards, landscaping trends, and innovations in paving materials and techniques.

It's crucial for contractors to master the complexities of paving and landscaping, blending design principles, technical know-how, and sustainability practices to construct spaces that are visually appealing, functional, and environmentally responsible. By doing so, they can contribute to the creation of enduring and meaningful landscapes and infrastructures.

In essence, delving deep into the subtleties of paving and landscaping, utilizing advanced materials and technologies, and embracing sustainability, can help contractors set new benchmarks in construction excellence. This multifaceted approach ensures that every project not only meets but exceeds, the expectations of users and stakeholders, reflecting a commitment to quality, innovation, and environmental stewardship.

Specialized Construction Techniques:

In special construction circumstances, unconventional techniques are pivotal. For instance, underwater construction demands caissons and cofferdams to create a dry work environment, whereas construction in extreme climates might require specialized insulating materials and construction methods to cope with extreme temperatures or weather conditions.

Underwater Construction:

Underwater construction employs techniques like the use of caissons, which are watertight structures, used for construction work below water level. Cofferdams are temporary enclosures built within or across bodies of water to allow the enclosed area to be pumped out, creating a dry work environment. The use of underwater concreting, achieved through tremie pipes, ensures that concrete can be placed underwater without washing out the cement content.

Construction in Extreme Climates:

In extremely cold climates, frost-protected shallow foundation designs help in preventing frost heave, while the choice of insulating materials like SIPs (Structural Insulated Panels) and ICFs (Insulating Concrete Forms) are critical. In contrast, hot climates necessitate reflective materials, cool roofs, and advanced HVAC systems to manage indoor temperatures effectively. Buildings in hurricane-prone areas often use reinforced concrete and steel structures, designed to withstand high winds and flying debris.

Risk Mitigation:

Understanding and mitigating risks in such specialized conditions involve comprehensive site analysis, advanced engineering solutions, and adherence to stringent safety protocols. For underwater construction, divers might undergo extensive training and utilize specialized equipment, and continuous monitoring of water conditions is vital to manage the risks of underwater currents and visibility issues.

For construction in extreme climates, meticulous planning around seasonal weather patterns, understanding the local microclimate, and implementing appropriate building designs and materials are necessary to mitigate the risks associated with extreme temperatures, heavy snow, or high winds.

Practical Application:

For instance, in the construction of offshore oil platforms, engineers and contractors have to employ specialized underwater construction techniques to build structures that can withstand the harsh marine environment, high pressures, and corrosive saltwater, ensuring the safety and longevity of the platform. In polar regions, buildings must be constructed to handle extreme cold, using materials and designs that can withstand freezing temperatures and ice accumulation, like elevated foundations to prevent heat loss to the frozen ground.

Expert Knowledge and Uncommon Strategies:

Sophisticated technologies such as ROVs (Remotely Operated Vehicles) are increasingly being used in underwater construction for inspection and construction tasks, reducing the risks to human divers. For extreme climates, advanced thermal modeling and passive design strategies can be employed to optimize energy efficiency and comfort.

Specialized construction techniques demand a fusion of advanced technology, innovative materials, and expert knowledge to mitigate inherent risks and ensure the success of construction projects in challenging environments. Mastery over these techniques enables contractors to expand their capabilities and undertake a diverse range of projects, pushing the boundaries of construction possibilities.

Safety:

Adherence to Occupational Safety and Health Administration (OSHA) regulations is imperative for maintaining safety and mitigating risks in the construction sector. Non-compliance not only jeopardizes safety but can also lead to substantial fines and legal repercussions.

Essential OSHA Regulations:

1. **Fall Protection:** OSHA mandates the use of guardrails, safety nets, or personal fall arrest systems to protect workers on surfaces with unprotected edges or sides 6 feet above a lower level.
2. **Hazard Communication:** Employers must classify hazards and communicate them to workers through proper labeling and access to Safety Data Sheets (SDS). Training workers on hazards related to chemicals is a requisite.
3. **Scaffolding:** Contractors must ensure that scaffolding is sound, rigid, and sufficient to carry its own weight and four times the maximum intended load without settling or displacement.
4. **Respiratory Protection:** In environments with insufficient oxygen or harmful dusts, fogs, smokes, mists, fumes, gases, vapors, or sprays, employers must provide appropriate respiratory protection equipment.
5. **Lockout/Tagout Procedures:** Implementation of proper lockout/tagout procedures is crucial to prevent the unexpected energization or startup of machinery and equipment during service or maintenance activities.

6. **Electrical Wiring Methods:** OSHA requires grounding of electrical equipment, proper use of extension cords, and proper installation and maintenance of all electrical systems to prevent electrocutions.

7. **Powered Industrial Trucks:** Contractors must ensure that each powered industrial truck operator is competent to operate a powered industrial truck safely, as demonstrated by the successful completion of training and evaluation.

Implications of Non-Compliance:
- Severe fines and penalties can be imposed.
- Temporary or permanent cessation of operations may be ordered.
- The reputation of the contractor can be significantly tarnished, leading to a loss of business opportunities.
- Legal ramifications may include lawsuits and legal fees.

Ensuring Adherence:
1. **Regular Training and Education:** Continual training and education on OSHA regulations and safety protocols for all employees are crucial. Regular safety meetings and on-site training can help in reinforcing safety practices.

2. **Safety Audits and Inspections:** Regular safety audits and inspections should be conducted to ensure that all operations are in compliance with OSHA regulations and to identify and rectify any potential safety hazards.

3. **Implementation of Safety Programs:** Contractors should implement comprehensive safety programs that include clear safety policies, emergency response procedures, and reporting and documentation protocols.

4. **Clear Communication:** Effective communication regarding safety protocols, hazard identifications, and emergency response procedures should be maintained with all on-site personnel.

5. **Use of Personal Protective Equipment (PPE):** Enforcing the use of appropriate PPE like helmets, gloves, safety shoes, goggles, etc., can significantly reduce the risk of workplace injuries.

6. **Documentation and Recordkeeping:** Keeping accurate records of training, inspections, incident reports, and safety meetings can provide proof of compliance and help in identifying areas for improvement.

Adhering to OSHA regulations and maintaining a proactive approach to safety, contractors can not only ensure the well-being of the workers but also avoid the severe repercussions associated with non-compliance.

Safety Procedures and Protocols:
Standard safety procedures and protocols are crucial for minimizing risks and ensuring the well-being of workers on construction sites. Contractors play a pivotal role in enforcing these measures, and non-compliance can lead to serious repercussions.
1. **Job Safety Analysis (JSA):** Contractors must conduct a JSA before commencing any task to identify potential hazards and determine preventive measures, ensuring tasks are performed safely.

2. **Use of Personal Protective Equipment (PPE):** Enforcing the use of proper PPE such as helmets, gloves, safety shoes, eye protection, and high-visibility clothing is mandatory to protect workers from potential hazards.
3. **Tool and Equipment Safety:** Regular inspection and maintenance of tools and equipment, along with proper training on their usage, are essential to prevent accidents and injuries.
4. **Safety Signage:** Clear and visible safety signs should be placed around the site to warn of potential hazards and instruct workers on necessary precautions.
5. **Hazard Communication:** All potential hazards, including chemical and physical, must be clearly communicated to all workers, and Safety Data Sheets (SDS) should be readily available.
6. **Emergency Exit Routes:** Clear and unobstructed emergency exit routes must be established and communicated to all personnel on-site.
7. **Safety Training:** Regular safety training sessions and drills are crucial to reinforce safety knowledge and procedures.

Importance & Contractor's Role:
- These protocols are imperative to prevent accidents, injuries, and fatalities.
- Contractors are responsible for implementing, monitoring, and enforcing safety protocols and ensuring all workers are adequately trained and equipped to comply.
- By fostering a safety-conscious environment, contractors can minimize downtime, legal liabilities, and boost worker morale and productivity.

Reinforcement Techniques:
1. **Regular Safety Meetings:** Conducting daily or weekly safety meetings to discuss potential hazards, safe work practices, and lessons learned from past incidents can reinforce safety measures.
2. **Positive Reinforcement:** Encouraging safe behavior through rewards and recognition can enhance compliance with safety procedures.
3. **Visible Leadership:** When management demonstrates a commitment to safety by being present on-site and adhering to safety protocols, it reinforces the importance of safety among workers.

First Aid and Emergency Response:
1. **Principles:**
   - **Preservation of Life:** Prioritize life-saving measures.
   - **Prevention of Further Harm:** Protect the injured from any more harm.
   - **Promotion of Recovery:** Provide aid to accelerate the healing process.
2. **Preparation:**
   - Contractors must ensure the availability of fully equipped first-aid kits at accessible locations.
   - Assign trained first-aid providers and ensure all workers know who they are and how to reach them.
   - Develop and communicate clear emergency response procedures, including emergency contact numbers and evacuation routes.

3. **Promptness and Effectiveness:**
   - Training in advanced first aid allows for quick and effective response to severe injuries, preventing further complications and promoting recovery.
   - Regular drills and training on emergency response can ensure the preparedness of workers and reduce response time in real emergencies.
4. **Advanced First Aid:** Knowing advanced first-aid procedures, like CPR and controlling severe bleeding, is crucial as it can be life-saving in the critical time before professional medical help arrives.

The adherence to safety procedures and the readiness for first aid and emergency response are indispensable components in construction projects, directly impacting the well-being of workers and the overall success of the project. Contractors, with their pivotal role, must be diligent in maintaining safety and preparedness at all times.

Safety Equipment and Gear:

Safety equipment and Personal Protective Equipment (PPE) are paramount in construction settings to mitigate the risk of injuries and fatalities. Here are the crucial elements and their applications:

1. **Hard Hats:** Designed to protect against impacts and falling objects.
2. **Safety Glasses/Goggles:** Shield eyes from flying debris, splashes, and intense light.
3. **High-Visibility Clothing:** Essential for enhancing visibility, especially in low-light conditions, to avoid collisions.
4. **Steel-Toed Boots:** Protect feet from crushing injuries due to falling objects.
5. **Hearing Protection:** Crucial in high-noise environments to prevent hearing damage.
6. **Respirators:** Filter airborne contaminants, crucial in areas with dust, mold, or harmful fumes.
7. **Harnesses:** Vital for fall protection when working at heights.

Repercussions of Neglect:

Neglecting to use safety gear can lead to serious injuries, fatalities, legal implications, increased insurance premiums, loss of productivity, and damaged reputation. Fines and penalties from regulatory bodies like OSHA can also be imposed for non-compliance.

Enforcement by Contractors:

1. **Regular Safety Training:** Educate workers on the correct use and importance of PPE and conduct refresher courses.
2. **Strict Enforcement:** Adopt a zero-tolerance policy for non-compliance with PPE requirements.
3. **Safety Audits:** Regularly inspect the workplace to ensure adherence to safety standards and correct use of PPE.
4. **Positive Reinforcement:** Recognize and reward adherence to safety protocols to encourage compliance.

Hazard Identification and Risk Management:

Identifying potential hazards and managing risks are pivotal for maintaining a safe construction site.

1. **Risk Assessment:** Regularly conduct risk assessments to identify and analyze potential hazards associated with each task, and implement control measures.
2. **Job Safety Analysis (JSA):** Break down each job into steps, identify potential hazards per step, and determine preventive measures.
3. **Safety Walkthroughs:** Regular inspections of the construction site can help in early detection of unsafe conditions or practices.
4. **Hazard Communication:** Clearly communicate identified hazards to all workers and ensure proper labeling of hazardous substances.
5. **Emergency Preparedness:** Develop and communicate clear emergency response plans, including evacuation routes and assembly points.

Proactive Safety Culture:
1. **Safety Leadership:** Demonstrating commitment to safety at all levels of the organization fosters a safety-conscious environment.
2. **Employee Engagement:** Involving workers in safety planning and decision-making enhances their commitment to safety practices.
3. **Continuous Learning:** Encourage learning from incidents and near misses and implement corrective actions to prevent reoccurrence.
4. **Safety Performance Monitoring:** Regularly review safety performance through safety metrics and Key Performance Indicators (KPIs) to identify areas for improvement.

Mitigating Impact of Identified Hazards:
1. **Elimination/Substitution:** Remove the hazard or replace hazardous processes/materials with safer alternatives.
2. **Engineering Controls:** Modify equipment or processes to reduce exposure to the hazard.
3. **Administrative Controls:** Change work practices and policies to reduce risk, including rotating workers and scheduling regular breaks.
4. **Use of PPE:** Provide appropriate personal protective equipment to shield workers from hazards.

Creating a proactive safety culture and rigorously adhering to safety equipment usage are paramount in mitigating risks and ensuring a safe and productive construction environment. These practices not only protect the well-being of the workers but also contribute significantly to the project's overall success and the organization's reputation.

Safety Training and Awareness:
An effective safety training program is paramount for enhancing construction site safety. Here are key components of such a program:
1. **Comprehensive Curriculum:** Covering all aspects of safety, including hazard recognition, PPE usage, emergency response, and first aid.
2. **Hands-on Training:** Practical demonstrations and drills to ensure workers can apply the knowledge acquired.
3. **Regular Updates and Refreshers:** Keeping abreast of the latest safety standards and regulations, and incorporating them into the training modules.

4. **Accessibility:** Delivering training in multiple formats, such as online, in-person, and printed materials, to accommodate varied learning preferences.
5. **Evaluation and Feedback:** Assessing participants' understanding and gathering feedback for continuous improvement of the training program.

Contribution to Safety:
1. **Enhanced Awareness:** Regular training fosters a safety-conscious work environment, making workers more vigilant about potential hazards.
2. **Reduced Incidents:** Better-informed and trained workers are less likely to make errors leading to accidents.
3. **Improved Compliance:** Awareness of safety regulations ensures adherence, reducing the likelihood of legal complications and penalties.
4. **Increased Productivity:** A safer work environment fosters worker morale and reduces downtime due to accidents.

Safety Documentation and Reporting:
Maintaining accurate safety documentation is crucial for verifying compliance with safety regulations and for analyzing safety performance. Here's how it plays a significant role:
1. **Legal Compliance:** Accurate records are essential for demonstrating adherence to safety laws and for defending against legal claims.
2. **Incident Analysis:** Documenting incidents enables in-depth analysis to identify root causes and implement corrective actions.
3. **Performance Monitoring:** Safety records aid in monitoring safety performance over time and in identifying trends and areas for improvement.
4. **Transparency and Accountability:** Comprehensive documentation fosters a sense of responsibility among workers and promotes transparency in safety operations.

Regular Safety Audits and Inspections:
1. **Proactive Identification of Hazards:** Regular audits reveal unsafe conditions and practices before they lead to accidents.
2. **Enhanced Compliance:** Inspections ensure ongoing adherence to safety standards and regulations, reducing the risk of non-compliance penalties.
3. **Continuous Improvement:** Identifying areas for improvement during audits enables the implementation of enhanced safety measures and protocols.
4. **Employee Engagement:** Inviting workers to participate in safety audits fosters a sense of ownership and commitment to safety practices.

Maintaining meticulous safety documentation and conducting regular audits and training are pivotal for creating and maintaining a safe construction environment, thereby safeguarding the well-being of the workforce and ensuring the seamless progress of the project.

Importance of Emergency Evacuation Plans:
Emergency evacuation plans are indispensable in construction settings, ensuring a swift and organized response to emergencies, minimizing harm, and saving lives. They are critical due to the varied and often hazardous nature of construction environments, where rapidly evolving situations can pose significant risks to personnel.

Components of Effective Emergency Evacuation Plans:
1. **Clear Egress Paths:** Designating and maintaining unobstructed routes to exit points.
2. **Designated Assembly Areas:** Identifying safe locations where personnel should gather post-evacuation.
3. **Role Assignments:** Assigning specific tasks such as roll call, first aid, and liaison with emergency services to designated individuals.
4. **Alarm Systems:** Implementing effective and distinct alarm systems to alert personnel to different types of emergencies.
5. **Regular Drills:** Conducting periodic evacuation drills to reinforce procedures and identify areas for improvement.
6. **Signage and Information:** Installing clear signage indicating evacuation routes and providing information on emergency procedures.
7. **Emergency Equipment:** Placing emergency equipment like fire extinguishers and first aid kits in accessible locations.
8. **Special Needs Consideration:** Developing plans to assist personnel with mobility or other special needs during an evacuation.

Ensuring Familiarity and Execution:
1. **Regular Training:** Providing frequent training sessions to ensure all personnel, including new hires, understand the evacuation procedures.
2. **Drills and Simulations:** Carrying out unannounced drills to simulate real emergency conditions and assess the readiness of the personnel.
3. **Feedback and Improvement:** Collecting feedback post-drills to address concerns and modify the evacuation plan as needed.
4. **Visible Information:** Displaying evacuation procedures prominently in multiple locations to reinforce knowledge.
5. **Individual Accountability:** Ensuring each worker understands their responsibility to be aware of and comply with evacuation procedures.
6. **Communication Strategies:** Establishing clear communication channels to relay evacuation instructions swiftly and clearly during an emergency.

By developing, implementing, and reinforcing emergency evacuation plans, contractors can significantly mitigate the risks associated with emergencies on construction sites, ensuring the safety of all personnel.

Project Management:

In construction projects, planning and scheduling are foundational components to assure projects are completed on time, within scope, and budget. Contractors harness various techniques and tools to structure project timelines effectively.

Principles of Planning and Scheduling:

Effective planning requires a meticulous breakdown of project activities, resources needed, and the time required for each task. Contractors must coordinate with various stakeholders to gather input on task durations and dependencies. Scheduling, then, involves allocating resources to these planned activities and sequencing them in an optimal manner to avoid delays.

Use of Gantt Charts:

Gantt charts are pivotal scheduling tools that represent the start and finish dates of the project's elements. They display the dependencies between activities and allow contractors to visualize the project timeline and monitor progress effectively. Gantt charts are invaluable in identifying bottlenecks and realigning resources to maintain the project schedule.

Critical Path Method (CPM):

CPM is a systematic approach that identifies the sequence of tasks that represent the longest path through the project, thus defining the shortest possible project duration. By pinpointing critical tasks that cannot be delayed without affecting the project's end date, contractors can allocate resources strategically to ensure that the critical path is maintained, and delays are avoided.

Implications of Improper Planning:

Inadequate planning and scheduling can lead to extensive delays, cost overruns, resource conflicts, and ultimately, project failure. Disorganized timelines can result in inefficient resource utilization and can increase the likelihood of disputes amongst project stakeholders due to unmet expectations and miscommunications.

Contingency Planning:

Contingency planning is crucial to mitigate the impacts of potential delays. It involves identifying potential risks and developing proactive strategies to address them. By having predefined responses to common issues, contractors can rapidly address problems as they arise, minimizing the impact on the project schedule. It often includes allocating buffer times within the schedule to accommodate unforeseen delays or disruptions.

For instance, if a construction project involves building a high-rise building, any delay in laying the foundation will have a cascading effect on all subsequent activities. Utilizing Gantt charts and CPM, the contractor can visualize these dependencies and allocate resources optimally to ensure that the foundation is completed on time, and any potential delays are addressed promptly through contingency plans.

The amalgamation of diligent planning and robust scheduling, empowered by tools like Gantt charts and the Critical Path Method, is quintessential for construction project success. It not only fortifies the project against unforeseen challenges but also serves as a linchpin for effective resource allocation, cost management, and stakeholder satisfaction. The sagacious contractor, versed in these techniques, can navigate the multifaceted landscape of construction projects, ensuring their seamless execution and triumphant completion.

Cost Management and Control:

In the realm of construction projects, strategic cost management and control are paramount. The meticulous interplay of cost estimation, budgeting, and value engineering plays a pivotal role in optimizing project costs.

Cost Estimation:

Cost estimation is the calculated approximation of the costs associated with the project's activities. Contractors employ various methods such as unit cost estimating, parametric estimating, and detailed estimating to assess the costs of labor, materials, equipment, and overhead. Leveraging accurate cost estimation methods is crucial to avoid underestimations that can lead to budget overruns.

Budgeting:

Once the costs are estimated, contractors delineate the budget, allocating funds to different project components, and establishing cost baselines. Continuous monitoring is vital to ensure that the expenditures are aligned with the planned budget, and any variances are identified and addressed promptly.

Value Engineering:

Value engineering is a systematic method to enhance the value of goods, products, or services. In construction, it involves analyzing design elements and functions to optimize the project's cost and performance. By exploring alternative construction methods, materials, and design solutions, contractors can achieve more cost-effective and value-oriented outcomes.

Repercussions of Cost Overruns:

Cost overruns can have severe implications, such as reduced profitability, strained relationships with stakeholders, litigation, and even project failure. They can tarnish the reputation of the contractor and may lead to loss of future business opportunities.

Strategies to Avoid Financial Pitfalls:

1. **Precise Estimation:** Develop detailed and accurate cost estimates, considering all potential expenditures and contingencies.
2. **Regular Monitoring:** Continuously monitor actual costs against the budget, identifying variances early and implementing corrective actions promptly.
3. **Risk Management:** Identify, assess, and mitigate financial risks proactively through comprehensive risk management plans.
4. **Contract Clarity:** Ensure clear and concise contracts to avoid disputes and ensure a mutual understanding of financial obligations.
5. **Change Order Management:** Effectively manage changes in project scope, cost, and schedule to avoid unnecessary expenses.

Consider a scenario where a contractor is constructing a commercial building. If the contractor does not conduct accurate cost estimation and fails to allocate sufficient funds for high-quality materials, it can lead to suboptimal construction quality and potentially, structural failures, thus emphasizing the need for accurate cost estimation, stringent budgeting, and value-optimized engineering solutions.

Mastering the dynamics of cost management and control, infused with precise estimation, prudent budgeting, and innovative value engineering, is pivotal for contractors aiming to navigate the fiscal intricacies of construction projects successfully. By embracing these strategies and fostering a culture of financial diligence, contractors can circumvent financial quandaries and steer their projects toward a prosperous completion. The sagacity in fiscal management not only sustains project viability but also underpins the integrity and reputation of the contracting entity in the competitive construction landscape.

Quality Management and Control:

Essential Components:

Quality Management in construction is rooted in four key components: Quality Planning, Quality Assurance (QA), Quality Control (QC), and Quality Improvement.

1. **Quality Planning:**
   - Establishing quality objectives and requirements.
   - Developing quality plans and documents to ensure compliance with quality standards and specifications.

2. **Quality Assurance:**
   - Systematic activities and management systems to ensure the project will meet the defined quality standards.
   - It encompasses process-oriented activities, monitoring of processes, and preventive actions to avoid defects.

3. **Quality Control:**
   - Inspection and verification of completed activities and elements to ensure adherence to project quality requirements.
   - Identifying and addressing discrepancies through corrective actions.

4. **Quality Improvement:**
   - Continuous efforts to enhance construction processes, reduce waste, and increase overall project value.

Implementation of QA and QC:

Contractors implement QA by developing a structured system to manage and improve processes, with the adoption of international standards like ISO 9001 being prevalent. Comprehensive quality management plans, regular audits, and process reviews are integral to ensure adherence to quality standards and continual improvement.

QC is implemented through stringent inspections and testing of materials and workmanship, ensuring every construction element meets the predefined quality criteria. Utilizing advanced technology and tools, contractors can accurately monitor and measure quality levels.

Impact of Poor Quality Management:

Poor quality management can lead to severe, multifaceted repercussions, including structural failures, costly rework, legal liabilities, and a tarnished reputation. The fallout can extend to loss of business, strained client relationships, and potential financial ruin.

Preventative Measures:

1. **Proactive Planning:** Detailed quality plans outlining standards, inspection criteria, and acceptance levels.
2. **Regular Training:** Periodic training sessions for workforce on quality standards and best practices.
3. **Robust Documentation:** Maintaining comprehensive records of quality activities, inspections, tests, and modifications.
4. **Advanced Technology:** Utilizing modern tools and software to monitor quality in real-time and identify discrepancies early.

5. **Client Engagement:** Continuous communication with clients to understand their expectations and get feedback on delivered quality.

Quality Management and Control are paramount in construction, with meticulous planning, assurance, control, and improvement being the cornerstones. By implementing robust QA and QC practices, contractors can ensure the delivery of projects that align with client expectations and stringent quality standards, mitigating risks and enhancing overall project value. Recognizing the profound implications of subpar quality, embracing preventative measures, and continuous improvement can sustain project success and elevate the contractor's standing in the construction sector.

Contract Administration:
Role in Construction Project Management:
Contract Administration is pivotal in managing the agreement between the contractor and the client, ensuring that the contractual obligations are fulfilled by all parties involved. It revolves around the enforcement, interpretation, and execution of the contract's terms and conditions and oversees compliance with the agreed specifications, quality, and timelines.

Ensuring Contract Provisions are Met:
Contractors deploy several strategies to ensure the fulfillment of contract provisions:
1. **Comprehensive Understanding:** A thorough understanding of contractual terms is crucial to ensure compliance.
2. **Monitoring and Reporting:** Regular monitoring of work progress and meticulous reporting ensure that the contractual conditions regarding project timelines and standards are met.
3. **Effective Communication:** Continuous liaison with stakeholders, suppliers, and subcontractors is essential to align everyone with the contractual obligations and address any arising issues promptly.

Handling Contract Modifications and Disputes:
1. **Contract Modifications:**
   - Any changes or modifications in the contract are carefully documented in the form of contract amendments, which detail the revised scope, cost, and time implications.
   - Contractors employ meticulous documentation and clear communication to ensure mutual agreement on any contract modifications.
2. **Dispute Resolution:**
   - Contract disputes are preferably resolved through negotiations and mediations to avoid costly and time-consuming legal battles.
   - Alternative Dispute Resolution (ADR) methods like arbitration can also be employed when mutual agreement seems unattainable.

Importance of Clear Contractual Terms:
Clear, concise, and unambiguous contractual terms are the bedrock of risk mitigation in construction projects. They:

1. **Minimize Misinterpretations:** Clear terms reduce the chance of misunderstandings and disputes.
2. **Define Obligations & Responsibilities:** They delineate the obligations, rights, and responsibilities of each party, ensuring alignment of expectations.
3. **Provide Legal Safeguard:** They act as a legal safeguard against non-compliance, safeguarding the interests of the parties involved.
4. **Outline Risk Allocation:** Clear contractual terms specify how risks are allocated between the contractual parties, thus helping in managing financial and legal exposures efficiently.

Contract Administration plays a vital role in construction project management by overseeing the fulfillment of contractual obligations and managing any amendments or disputes that arise during the project lifecycle. The emphasis on clear contractual terms and proactive management of contracts is instrumental in mitigating legal and financial risks, ensuring smooth project execution, and maintaining harmonious relationships between all project stakeholders. Through effective contract administration, contractors can navigate the complexities of construction agreements, ensuring the mutual satisfaction of all parties involved and the successful delivery of construction projects.

Planning and Scheduling:
Detailed and meticulous planning is the cornerstone, involving the thorough breakdown of project tasks, allocation of resources, and scheduling. Utilizing techniques such as the Critical Path Method (CPM) and employing Gantt charts allow contractors to visualize project timelines and allocate resources effectively, reducing the risk of overruns. Delays can be devastating; hence, the inclusion of buffer times and contingency plans is critical.

Cost Management and Control:
Astute cost management and control are non-negotiable. Cost estimation, budgeting, and value engineering are pivotal processes to optimize project costs. The meticulous scrutiny of project expenses, regular budget reviews, and proactive adjustments are instrumental in averting financial pitfalls. Cost overruns can have severe repercussions, leading to strained client relationships and diminishing profits.

Quality Management and Control:
Implementing stringent quality assurance and quality control processes is essential. This involves systematic activities, regular inspections, and testing to ensure the project meets predefined standards and client expectations. Failing in quality management can lead to rework, increased costs, and reputational damage. Adopting a proactive approach to identify and address quality issues early is crucial.

Contract Administration:
Contract administration is the linchpin, focusing on ensuring that contract provisions are met. Clear contractual terms, regular reviews, and effective handling of modifications and disputes are crucial to mitigate legal and financial risks. Misinterpretations or violations of contract terms can lead to disputes, delays, and additional costs.

Integration and Interrelation:

Every component mentioned is intertwined. Effective planning and scheduling provide a framework for cost management, while quality management ensures adherence to standards, reducing the risk of costly rework and delays. Contract administration acts as the governing body, ensuring legal and contractual compliance throughout the project lifecycle.

Common Challenges:

Contractors often grapple with uncertainties and unpredictabilities, such as unforeseen site conditions, fluctuations in material prices, and changes in client requirements, all requiring agile and responsive project management to adjust plans, budgets, schedules, and resources promptly.

Strategies for Success:

1. **Robust Risk Management:** Identifying, assessing, and mitigating risks early can prevent many common project challenges.
2. **Regular Monitoring and Review:** Continual oversight of project progress against the plan, coupled with regular reviews of cost, quality, and schedules, enables early detection and resolution of issues.
3. **Effective Communication:** Regular, clear, and concise communication with all stakeholders is essential to avoid misunderstandings and ensure everyone is aligned on project goals and expectations.
4. **Investment in Technology:** Leveraging advanced project management software and technologies can enhance planning, monitoring, and control, increasing overall project efficiency and success.
5. **Continual Learning and Improvement:** Regularly updating skills, adopting new methodologies, and learning from previous projects contribute to enhanced project management capabilities and outcomes.

In real-world applications, successful contractors seamlessly integrate these principles to navigate the complexities of construction projects, from small residential builds to sprawling commercial developments, ensuring project delivery that is on time, within budget, and to the required quality standards. Adopting best practices and leveraging deep knowledge in these areas significantly contributes to achieving project and business success, aligning the interests of clients, contractors, and all other stakeholders.

Remember, these insights are tailored to provide foundational knowledge in project management as part of preparing for the Contractor License Exam, encapsulating the depth and breadth of knowledge required for aspiring contractors to excel in their endeavors.

Trade-Specific Knowledge:

In the electrical trade, the methods and techniques employed are paramount in ensuring that installations, repairs, and maintenance are executed proficiently, ensuring the safety and efficacy of electrical systems in construction projects. Here's a deeper look into some of these critical methods and techniques and their implications:

1. Wiring Methodologies:
- **Conduit Wiring:** Utilizing metal or plastic tubes to protect wires, conduit wiring is fundamental for preventing electrical hazards, particularly in exposed locations.

- **Cable Wiring:** Involves the use of armored or unarmored cables, and it's essential for providing insulated and protective wiring solutions in varied environments.

2. Circuit Design:
- **Branch Circuits:** Designed to deliver electrical energy to an outlet or a series of connected outlets, proper design is crucial to avoid overloading and potential fire hazards.
- **Feeder Circuits:** These serve multiple branch circuits and require meticulous calculation and design to ascertain the load and prevent system failures.
- **Dedicated Circuits:** Essential for appliances with high energy needs, proper implementation avoids overloads and disruptions in power supply.

3. Installation Practices:
- **Correct Phase Sequencing:** Ensuring the correct order of phases during installation is pivotal to prevent damage to equipment and ensure system reliability.
- **Grounding:** Proper grounding is fundamental for preventing electrical shocks and enhancing the safety of electrical systems.
- **Secure Mounting:** Firm and secure mounting of electrical components is vital to avoid loosening and subsequent failures.

Importance in Construction Projects:
- **Safety:** Proper wiring, circuit design, and installation practices are pivotal in minimizing risks of electrical shocks, fires, and other hazards.
- **Reliability:** Methodological and meticulous application of these techniques assures the uninterrupted and dependable operation of electrical systems.
- **Compliance:** Adherence to established methods and best practices ensures compliance with local and national electrical codes, avoiding legal repercussions and potential fines.
- **Cost-Efficiency:** Proper implementation of these methods reduces the occurrence of electrical failures, reducing maintenance and repair costs over time.

Practical Implications:
- **Optimized Power Distribution:** Effective circuit design and wiring methodologies ensure optimal power distribution, meeting the varying power demands of different construction elements efficiently.
- **Enhanced Equipment Lifespan:** Proper installation and maintenance practices extend the life expectancy of electrical components and equipment.
- **Sustainability:** Efficient electrical systems contribute to energy conservation and sustainability in construction projects.

So by applying precise methods and techniques in wiring, circuit design, and installation practices, electrical contractors play a critical role in shaping the safety, reliability, and efficiency of electrical systems in construction projects. These practices are not mere procedural formalities but are pivotal foundations ensuring the seamless integration and operation of electrical components within the complex framework of construction environments.

In the construction industry, particularly within the electrical trade, compliance with laws and codes is non-negotiable and is strictly enforced to uphold the safety and reliability of electrical systems. Here's a succinct exploration of some key electrical codes and laws:

National Electrical Code (NEC):
- **Objective:** The NEC sets the foundation for electrical safety in residential, commercial, and industrial occupancies, providing guidelines for electrical wiring, overcurrent protection, grounding, and installation of equipment.
- **Significance:** It mitigates the risk of electrical fires, shocks, and other hazards, ensuring the safe installation and operation of electrical equipment.

Occupational Safety and Health Administration (OSHA) Standards:
- **Objective:** OSHA has specific standards for the construction industry, addressing electrical safety requirements to protect employees from electrical hazards.
- **Significance:** Compliance with OSHA standards is crucial to ensure a safe working environment, reduce the risk of accidents, and avoid hefty fines and legal actions.

Local Building Codes:
- **Objective:** Local jurisdictions may have additional requirements and amendments to the NEC. Contractors must adhere to these local codes when performing electrical work.
- **Significance:** Local codes often address unique regional needs and conditions, ensuring that electrical installations are suitable and safe for the local environment.

Implications of Non-Compliance:
- **Legal Repercussions:** Failure to comply can lead to severe legal consequences, including fines, suspension of contractor licenses, and even imprisonment in cases of gross negligence leading to harm.
- **Financial Implications:** Non-compliance can result in costly repairs, retrofitting, and potential litigation, significantly impacting the financial stability of contractors.
- **Reputation Damage:** Violation of codes and laws can tarnish the reputation of contractors, leading to loss of clients and business opportunities.

Compliance Strategies:
- **Continued Education:** Regular training and staying updated on changes in laws and codes are essential for maintaining compliance.
- **Comprehensive Planning:** Thoroughly reviewing project requirements against applicable codes and laws during the planning phase mitigates the risk of non-compliance.
- **Consultation with Authorities:** Proactively engaging with local building officials and inspectors can clarify any ambiguities related to code compliance and prevent violations.

Real-World Application:
In practical terms, adherence to these laws and codes means meticulous attention to every detail in the electrical installation process, from wire sizing to placement of outlets, to ensure not only the functional reliability of electrical systems but, more critically, the safety of occupants and operators interacting with those systems.

Adherence to electrical codes and laws is an essential practice in safeguarding against the inherent risks associated with electricity. It not only solidifies the safety and reliability of electrical systems but also substantiates the integrity and professionalism of contractors in the industry. These rules are not mere bureaucratic red tape; they are lifelines ensuring the harmonious integration of electrical components within the living and working spaces of our communities.

Compliance with key electrical codes and laws is paramount to the safety and reliability of electrical systems within any construction project.

Electrical Codes and Laws:
1. **National Electrical Code (NEC):**
   - **Purpose:** It provides guidelines for electrical wiring, equipment installations, and overcurrent protection, setting the standard for safe electrical design, installation, and inspection.
   - **Safety & Reliability:** Ensures safe installations by mitigating risks of fire and electrical shocks, specifying requirements for electrical system reliability.
2. **Occupational Safety and Health Administration (OSHA) Regulations:**
   - **Purpose:** OSHA's standards protect workers from electrical hazards, outlining safety requirements for installation, maintenance, and use of electrical equipment.
   - **Safety & Reliability:** It helps in establishing a safe workplace by requiring protective equipment, insulating materials, and safe work practices, ultimately reducing electrical accidents and enhancing reliability.
3. **Local Electrical Codes:**
   - **Purpose:** Based on the NEC, local jurisdictions may have additional requirements or modifications to adapt to local needs and conditions.
   - **Safety & Reliability:** Addressing regional requirements, these codes ensure that electrical installations are optimal and safe for the specific locality, considering environmental and structural factors.

Ensuring Safety & Reliability:
- **Regulatory Compliance:** These laws and codes establish minimum standards for safe electrical practices and systems, preventing unsafe conditions leading to accidents and fires.
- **Systematic Inspection:** Regular inspections ensure that every aspect of electrical installations complies with codes, ensuring long-term safety and reliability.
- **Equipment Specifications:** Adherence to codes ensures the use of standardized, tested equipment, reducing risks of equipment failure and associated hazards.

Implications of Non-Compliance:
1. **Legal Consequences:** Violations can lead to severe penalties, including fines and license revocations, impacting the legal standing of the contractor.
2. **Financial Repercussions:** The costs associated with rectifications, legal fees, and fines can be substantial, impacting the contractor's financial health.
3. **Reputational Damage:** Non-compliance tarnishes the reputation of contractors, potentially resulting in a loss of business opportunities and clientele.

4. **Safety Hazards:** Above all, non-compliance risks the safety of individuals, leading to potential accidents, injuries, and loss of life.

Best Practices for Compliance:
1. **Regular Training & Updates:** Continuous learning and staying abreast of updates in laws and codes are crucial.
2. **Pre-Construction Planning:** Detailed review and planning in accordance with the codes prevent violations and subsequent repercussions.
3. **Consultation and Collaboration:** Proactive engagement with local code enforcement officers and electrical inspectors ensures alignment with all requirements and swift resolution of any issues.

Strict adherence to electrical codes and laws is non-negotiable for contractors who are responsible for the lifeblood of modern buildings—electrical systems. This adherence solidifies the safety, reliability, and integrity of electrical installations, reflecting a contractor's commitment to professional excellence and ethical conduct in safeguarding lives and properties.

Special Considerations in Electrical Work:
1. **High Voltage Systems:**
   - **Consideration:** Specialized training and equipment are essential when working with high-voltage systems due to the increased risk of electric shock, arc flash, and fire.
   - **Techniques:** Use of insulated tools, personal protective equipment (PPE), and adhering to stringent lockout/tagout procedures are critical to manage risks associated with high-voltage systems.
   - **Application:** Often used in industrial applications, utility services, and large commercial buildings where the demand for electrical power is high.
2. **Smart Technologies and Automation:**
   - **Consideration:** Incorporating smart technologies requires a comprehensive understanding of IoT devices, sensors, and automation protocols to ensure seamless integration and operation.
   - **Techniques:** Installation of advanced control systems, programming of automation logic, and ensuring cybersecurity of connected devices are fundamental aspects.
   - **Application:** Applied extensively in modern building projects for energy management, security, lighting, and HVAC control, enhancing building efficiency, and user convenience.
3. **Green Energy Solutions:**
   - **Consideration:** Implementing sustainable electrical solutions, such as solar panels and wind turbines, necessitates knowledge of renewable energy technologies and grid integration.
   - **Techniques:** Proper sizing, positioning, and installation of renewable energy systems and integration with existing electrical infrastructure are crucial.

- **Application:** Used in eco-friendly projects aiming to reduce carbon footprint and energy consumption, and it's often incentivized by government programs and policies.
4. **Energy Storage Solutions:**
   - **Consideration:** Incorporating energy storage solutions like batteries demands understanding of storage capacities, discharge rates, and integration with renewable energy sources.
   - **Techniques:** Correct installation, maintenance, and monitoring of energy storage systems are key to ensure optimal performance and safety.
   - **Application:** Vital in projects where energy reliability and availability are critical, and it aids in managing energy costs and ensuring uninterrupted power supply.

Advanced Techniques:
1. **Building Automation Systems (BAS):**
   - **Principle:** Utilizing advanced sensors and controls to automate building operations such as lighting, HVAC, and security.
   - **Implementation:** Careful planning, installation, and programming are required to optimize building operations and ensure user comfort and safety.
2. **Advanced Cable Management:**
   - **Principle:** Organizing and securing electrical wires and cables efficiently to prevent accidents and facilitate maintenance.
   - **Implementation:** Utilization of cable trays, conduits, and labeling, considering future expansions and modifications.
3. **Integrated Systems:**
   - **Principle:** Merging various building systems like security, fire alarm, and HVAC into a single, coherent system.
   - **Implementation:** Coordinated design and installation, ensuring compatibility and seamless interaction between different systems.

Special considerations and advanced techniques in electrical work are pivotal in contemporary construction projects. Mastery of high-voltage systems, smart technologies, green energy solutions, and advanced integration techniques positions contractors to meet the evolving demands of the construction industry, ensuring project success, sustainability, and user satisfaction. The detailed understanding and application of these specialized considerations and techniques are crucial for contractors to stay abreast of industry advancements and to effectively meet the dynamic needs of clients in today's technologically driven and environmentally conscious construction landscape.

Green Building Practices in Electrical Trade:
1. **Renewable Energy Sources:**
   - **Application:** The implementation of solar panels, wind turbines, and other renewable energy sources is pivotal.

- **Contractor Role:** Contractors need to comprehend the integration, installation, and maintenance of renewable systems, ensuring they comply with local codes and standards.
- **Efficiency:** Utilizing renewables can drastically reduce the dependency on non-renewable energy sources, lowering operational costs and carbon footprints.

2. **Energy-Efficient Appliances and Lighting:**
   - **Application:** Prioritize the use of Energy Star-rated appliances and LED lighting solutions, which consume less energy.
   - **Contractor Role:** Contractors should be adept in recommending and installing energy-efficient solutions, focusing on minimizing energy consumption and enhancing sustainability.
   - **Efficiency:** These solutions can reduce energy consumption by up to 75% compared to traditional counterparts, providing long-term cost savings.

3. **Energy Management Systems:**
   - **Application:** Implementing advanced energy management systems to monitor and control energy consumption in real-time.
   - **Contractor Role:** Understanding the installation and operation of these systems is crucial for contractors to facilitate energy conservation and optimal energy usage.
   - **Efficiency:** Such systems can provide significant energy savings by optimizing energy consumption based on demand and occupancy, and identifying energy wastage points.

Energy-Efficient Design and Construction:
1. **Building Envelope Optimization:**
   - **Application:** Optimal insulation, window placements, and the use of reflective materials can minimize heat gains and losses.
   - **Contractor Role:** Electrical contractors collaborate with design teams to optimize the building envelope and incorporate energy-efficient electrical systems and lighting design.
   - **Efficiency:** Properly designed and constructed building envelopes can significantly reduce HVAC loads and energy consumption.

2. **Smart Controls and Automation:**
   - **Application:** Smart thermostats, occupancy sensors, and intelligent lighting controls can significantly reduce energy consumption.
   - **Contractor Role:** Contractors must be proficient in installing and programming smart controls and should understand the interplay between various building systems.
   - **Efficiency:** Smart controls can adjust the operation of appliances and systems based on occupancy and preferences, leading to energy conservation.

3. **High-Efficiency HVAC Systems:**
   - **Application:** Incorporating high-efficiency HVAC systems and utilizing variable frequency drives can optimize energy use.

- **Contractor Role:** Contractors work closely with HVAC technicians to ensure seamless integration of electrical and mechanical systems.
- **Efficiency:** High-efficiency systems and drives can optimize energy use based on demand, reducing energy wastage and operational costs.

Advancements and Insights:
1. **Innovations in Green Building:**
   - Advancements like building-integrated photovoltaics (BIPV) and energy storage solutions are revolutionizing green building practices.
   - Utilization of innovative materials and technologies can enhance building sustainability and resilience.
2. **Importance of Continuing Education:**
   - Keeping abreast of the latest trends, technologies, and best practices in green building is essential for electrical contractors.
   - Engaging in professional development and earning relevant certifications can enhance contractors' expertise and marketability.
- For instance, contractors implementing green building practices in commercial buildings can lead to LEED certification, which not only reduces operational costs but also enhances the building's market value and appeal.

Integrating green building practices and energy efficiency in the electrical trade is not only a moral imperative but also a practical one. It's about blending innovation with responsibility to shape a sustainable future. A well-versed and certified contractor is instrumental in steering this change, becoming the harbinger of a new era in sustainable construction. Whether it's about harnessing the power of the sun or optimizing a building's energy use, every step counts towards creating a more sustainable and energy-efficient world.

Pivotal Methods and Techniques in Plumbing:
1. **Pipe Installation:**
   - **Precision and Methodology:** Proper measurement and cutting are crucial to avoid any misalignment or improper fitting, ensuring a seamless flow and reducing chances of leakage.
   - **Material Knowledge:** Understanding the appropriate materials, such as PVC, copper, or PEX, is essential, as incorrect materials can lead to premature failure or contamination.
   - **Slope Installation:** Correct slope installation is critical for drain pipes to avoid standing water, which can lead to corrosion and blockages.
2. **Leak Detection:**
   - **Pressure Testing:** This method involves isolating sections of the plumbing system and pressurizing them with air or water to identify leaks.
   - **Infrared Technology:** This advanced technique utilizes infrared cameras to detect temperature variations in walls and floors, pinpointing hidden leaks.

- **Impact:** Early detection and repair of leaks are paramount to prevent water damage, mold growth, and water wastage, maintaining the system's efficiency and reliability.
3. **Repair Methodologies:**
   - **Pipe Repair Clamps:** Utilized for temporary fixes on small leaks, these clamps can prevent water damage until a permanent repair is conducted.
   - **Pipe Relining:** This technique creates a "pipe within a pipe" to seal leaks and improve flow, offering a less invasive alternative to pipe replacement.
   - **Precision:** Accurate identification of the problem area and choosing the right repair method is crucial to avoid further damage and ensure the longevity of the plumbing system.

Impact on Plumbing System's Efficiency and Reliability:
1. **Precision and Accuracy:**
   - The meticulous implementation of methods and techniques is indispensable to avoid system failures, ensuring the durability and optimum performance of the plumbing system.
   - Precision in installation and repairs guarantees the correct functioning of valves, seals, and joints, which are critical for the system's leak-proof nature.
2. **Compliance with Codes and Standards:**
   - Adherence to plumbing codes and standards is mandatory to ensure the safety and reliability of the plumbing system.
   - Correct installation and repair methodologies ensure that the plumbing system operates efficiently, reducing the risk of contamination and water wastage.
3. **Economic and Environmental Impact:**
   - Precision and accuracy in plumbing techniques significantly reduce the likelihood of leaks and failures, which can lead to substantial water savings, lower repair costs, and conservation of natural resources.
   - Efficient and reliable plumbing systems contribute to the overall sustainability of buildings, promoting water conservation and energy efficiency.

Consider a multi-story building project; precision in pipe installation will dictate the efficiency of water distribution to each level, impacting the system's reliability and the building's overall water usage. Accurate leak detection and timely repair can prevent structural damage and unnecessary water loss, maintaining the integrity and sustainability of the building. Balancing precision, accuracy, and adherence to standards in plumbing methodologies is the backbone of creating dependable and efficient plumbing systems.

Essential Plumbing Codes and Laws:
1. **Backflow Prevention:**
   - **Code Specifications:** Plumbing codes necessitate the installation of backflow prevention devices to avoid contamination of potable water supplies.
   - **Importance:** These are critical for protecting water supplies from the reversal of water flow, preventing contamination and ensuring clean water accessibility.

- **Compliance:** Regular testing and maintenance of backflow preventers are mandated to guarantee continuous protection against water supply contamination.
2. **Proper Sanitation:**
   - **Sanitary Regulations:** Plumbing laws stipulate proper waste disposal and venting systems to avoid harmful sewer gases' infiltration into buildings.
   - **Health and Safety:** Ensuring proper sanitation safeguards occupants' health by preventing exposure to harmful bacteria and pathogens and avoiding foul odors.
   - **Implementation:** Adherence to sanitary codes is verified through inspections, ensuring proper slope, venting, and waste disposal systems are in place.
3. **Water Conservation:**
   - **Conservation Codes:** Many regions have laws focusing on water-saving fixtures and appliances to reduce water consumption and waste.
   - **Sustainability:** Compliance with these codes contributes to environmental conservation, reduces strain on water supplies, and promotes energy efficiency.
   - **Enforcement:** Regular audits and certification of plumbing systems ensure adherence to water conservation standards, promoting long-term sustainability.

Safety Protocols for Plumbing:
1. **Handling and Disposal of Waste:**
   - **Safe Handling:** Proper procedures for handling and disposing of waste materials, including the utilization of appropriate containers, are imperative to avoid contamination and exposure to harmful substances.
   - **Hazardous Materials:** Special protocols are enforced for dealing with hazardous materials, ensuring their safe disposal and minimizing environmental impact.
2. **Use of Personal Protective Equipment (PPE):**
   - **Mandatory PPE:** Use of gloves, goggles, and other protective gear is obligatory to safeguard plumbers from exposure to harmful substances and potential injuries.
   - **Training and Compliance:** Regular safety training and strict adherence to PPE guidelines ensure worker safety and minimize the risk of accidents and occupational hazards.

Impact of Safety Protocols and Laws on Plumbing Systems:
- **Enhanced Safety:** Implementation of stringent safety protocols and compliance with plumbing laws ensures the well-being of both the occupants and the workers, preventing accidents and health issues.
- **System Reliability:** Adherence to plumbing codes guarantees the installation of reliable and efficient plumbing systems, reducing the likelihood of failures and ensuring proper functionality.
- **Legal and Financial Security:** Compliance with laws and safety protocols mitigates legal risks and potential financial liabilities arising from accidents, system failures, and non-compliance fines.

In a hospital setting, where sanitation and uninterrupted water supply are crucial, meticulous adherence to plumbing codes and rigorous implementation of safety protocols are paramount. Proper backflow prevention is vital to prevent contamination, and stringent safety measures are essential to protect workers dealing with hazardous waste materials and high-risk environments. The balanced integration of laws, codes, and safety measures ensures the seamless operation of plumbing systems in such critical settings, safeguarding public health and well-being.

Special Considerations and Advanced Techniques in Plumbing:
1. **High-Pressure Systems:**
    - **Consideration:** Managing high-pressure systems requires intricate knowledge and specialized techniques to avoid system failures and ensure safety.
    - **Implementation:** Utilization of pressure relief valves and robust piping materials are crucial to mitigate the risks associated with high pressure.
    - **Implication:** Proper handling and installation are paramount to avoid pipe bursts, leaks, and subsequent water damage, ensuring system longevity and reliability.
2. **Integration of Water-Saving Technologies:**
    - **Innovation:** The incorporation of advanced, water-efficient fixtures and appliances is pivotal in modern construction to conserve water.
    - **Technique:** Employing low-flow faucets, dual-flush toilets, and water-efficient appliances significantly reduces water consumption.
    - **Consideration:** Selection and installation of these technologies require thorough knowledge of product specifications and installation procedures to optimize performance.

Green Building Practices and Energy Efficiency in Plumbing:
1. **Water Conservation Techniques:**
    - **Practice:** Implementing greywater systems and rainwater harvesting contributes substantially to water conservation in construction projects.
    - **Efficiency:** These systems, coupled with water-efficient fixtures, optimize water usage, reducing waste and promoting sustainability.
    - **Impact:** Enhanced water efficiency contributes to reduced utility bills, promotes environmental conservation, and meets green building certification standards.
2. **Eco-friendly Materials:**
    - **Material Selection:** Utilizing lead-free pipes, fittings made from recycled materials, and non-toxic adhesives are fundamental in green plumbing.
    - **Sustainability:** The employment of eco-friendly materials minimizes environmental impact and enhances the sustainability of plumbing systems.
    - **Certification:** Adherence to green material standards is often necessary for obtaining green building certifications like LEED.
3. **Innovative Sustainable Solutions:**

- **Technological Advancement:** Integration of smart water management systems and leak detection technologies enhances water conservation and system reliability.
- **Application:** These innovations provide real-time monitoring and control over water usage, enabling immediate response to any irregularities and minimizing water waste.
- **Sustainable Impact:** Embracing such innovations promotes sustainability, resource conservation, and reduces the ecological footprint of buildings.

Consider a high-rise building project incorporating advanced plumbing techniques and green practices. Managing high-pressure systems is crucial to prevent failures in the extensive network of pipes running through the building. The integration of water-saving technologies and fixtures, coupled with the use of eco-friendly materials, is paramount to achieve water efficiency and sustainability. The incorporation of smart water management systems ensures optimal water use, immediate leak detection, and contributes to the overall sustainable profile of the building. The meticulous implementation of these advanced techniques and green practices ensures the project's success, longevity, and compliance with green building standards, showcasing the transformative potential of innovative and sustainable plumbing solutions in modern construction.

Trade-Specific Methods and Techniques in HVAC:
1. **System Design Principles:**
   - **Design Process:** Contractors use Manual J for load calculations to determine the appropriate size of HVAC units and Manual D for designing the correct ductwork to ensure proper air distribution.
   - **Optimization:** Zoning systems are often implemented, allowing different areas to maintain different temperatures, optimizing energy usage.
   - **Consideration:** Proper system design is crucial to avoid over or undersized systems, which can lead to inefficiency, increased wear and tear, and reduced comfort.
2. **Installation Procedures:**
   - **Procedure Accuracy:** Precision in installation, proper sealing of ductwork, and accurate refrigerant charging are essential to ensure the system's optimal performance and longevity.
   - **Quality Assurance:** Post-installation, systems undergo thorough testing and balancing to verify the airflow and ensure all components operate efficiently and safely.
   - **Standard Adherence:** Following the manufacturer's installation guidelines is imperative to avoid any discrepancies that can lead to system failure.
3. **Maintenance Practices:**
   - **Regular Maintenance:** Scheduled preventive maintenance includes cleaning, adjusting, and checking systems to detect and address any emerging issues promptly.

- **Efficiency:** Regular filter changes, coil cleaning, and duct maintenance are pivotal to maintain system efficiency and indoor air quality.
- **Longevity Impact:** Adherence to proper maintenance practices extends the system's lifespan, reduces the frequency of repairs, and ensures sustained performance.

Trade-Specific Laws and Codes in HVAC:

1. **HVAC Codes and Regulations:**
   - **Energy Efficiency:** Energy codes like the International Energy Conservation Code (IECC) dictate the minimum energy efficiency levels for HVAC systems.
   - **Environmental Protection:** Regulations like the Clean Air Act govern the use of refrigerants to mitigate environmental impact.
   - **Occupant Well-being:** Building codes stipulate the minimum requirements for ventilation and indoor air quality to ensure the well-being of the occupants.

2. **Compliance Implications:**
   - **Design and Installation:** Adherence to laws and codes is crucial from the design phase through installation to avoid legal repercussions and ensure the system's safety and efficiency.
   - **Certification:** Compliance with codes often involves obtaining necessary certifications and permits, demonstrating the system's legality and integrity.
   - **Maintenance and Repairs:** Regular inspections and maintenance are mandated to verify continued adherence to codes, detect deviations, and implement corrective actions.

Consider a commercial building project where HVAC contractors employ advanced design principles to optimize energy use, such as implementing zoning systems and adhering strictly to Manuals J and D during the design phase. Precision in installation and adherence to manufacturer's guidelines ensure optimal performance, while regular maintenance practices like scheduled cleaning and adjustments maintain system efficiency and indoor air quality. The stringent adherence to HVAC codes and regulations, from design to maintenance, not only ensures the system's compliance with energy and environmental standards but also safeguards the well-being of the occupants, illustrating the comprehensive and multifaceted role of HVAC contractors in the construction process.

Safety Protocols in HVAC Trade:

1. **Refrigerant Handling:**
   - **Proper Handling:** Contractors must follow EPA Section 608 guidelines to manage refrigerants correctly, avoiding leaks and exposures which can be harmful.
   - **Recovery and Disposal:** Refrigerant recovery, recycling, or reclaiming must be performed, and disposal must comply with legal and environmental regulations.
   - **Training and Certification:** Proper training and certification are required to handle refrigerants to ensure safety and compliance with environmental laws.

2. **Lifting Procedures:**

- **Equipment Handling:** Proper lifting techniques and mechanical aids should be used when handling heavy HVAC units and equipment to prevent musculoskeletal injuries.
- **Safety Training:** Regular training sessions on manual handling and lifting procedures ensure workers are knowledgeable about safe practices and potential risks.
- **Risk Assessment:** Identifying potential hazards and conducting risk assessments prior to lifting can significantly mitigate the risks of injuries.

3. **Use of Personal Protective Equipment (PPE):**
   - **Protective Gear:** Wearing appropriate PPE like gloves, safety glasses, and ear protection is crucial to protect against various hazards, such as cuts, debris, and noise.
   - **Respiratory Protection:** Respirators might be required when exposed to harmful dust, fumes, or other airborne contaminants.
   - **Consistency:** Regular use of PPE, combined with proper training and enforcement, is fundamental to maintaining a safe working environment.

Importance of Safety Protocols in HVAC Trade:
- **Mitigating Risks:** Strict adherence to safety protocols minimizes the risk of accidents and injuries, safeguarding the well-being of HVAC workers.
- **Legal Compliance:** Following established safety procedures and protocols ensures compliance with Occupational Safety and Health Administration (OSHA) regulations and other pertinent safety laws, avoiding legal ramifications.
- **Productivity and Morale:** A safe working environment contributes to higher worker morale and productivity by reducing downtime due to accidents and fostering a positive workplace culture.

In real-world scenarios, HVAC technicians often encounter diverse and potentially hazardous working conditions. For instance, when servicing a commercial HVAC system, a technician must carefully handle refrigerants, adhering strictly to EPA guidelines, and employ proper lifting procedures and mechanical aids when handling heavy equipment. Constant use of appropriate PPE, coupled with meticulous adherence to established safety protocols, is pivotal in navigating the myriad of risks present in HVAC work, ensuring both the safety of the technician and compliance with prevailing safety regulations. This disciplined approach to safety is not only a legal mandate but a moral and professional obligation, underscoring its instrumental role in the HVAC trade.

Special Considerations and Advanced Techniques in HVAC Work:
1. **Diverse Climate Conditions:**
   - **Climate-Specific Design:** HVAC systems must be designed to perform efficiently in the specific climatic conditions they will operate, whether it's extreme cold, heat, or humidity. This includes selecting appropriate refrigerants and designing systems to manage loads effectively.

- **Zoning Systems:** Creating different zones within a building can optimize energy use and ensure consistent comfort, allowing for adjustments based on the varying needs of different areas.
- **Climate-Resilient Equipment:** Using equipment and materials that can withstand extreme weather conditions is crucial in maintaining system longevity and reliability.

2. **Smart and Automated Climate Control Systems:**
   - **Building Automation Systems (BAS):** These systems integrate various building services, including HVAC, to optimize performance, comfort, and energy efficiency.
   - **IoT Devices:** The integration of Internet of Things (IoT) devices enables advanced monitoring and control of HVAC systems, contributing to energy savings and enhanced user comfort.
   - **Smart Thermostats:** These devices allow for remote monitoring and control of temperature settings, adapting to user preferences and optimizing energy use.

3. **Innovations in Heating and Cooling Technologies:**
   - **Variable Refrigerant Flow (VRF) Systems:** These advanced systems allow for varying refrigerant flow to different parts of a building, optimizing comfort and efficiency.
   - **Heat Pumps:** Advances in heat pump technology enable more efficient heating and cooling, even in extreme climatic conditions.
   - **Radiant Heating and Cooling:** These systems provide comfort by heating and cooling surfaces, offering a more efficient and comfortable alternative to traditional forced-air systems.

In a modern office building, HVAC contractors may employ a combination of these advanced techniques. A climate-specific design can ensure the system operates optimally in the local weather conditions, while zoning systems can address the varying needs of different spaces like conference rooms and open-plan offices. Integrating a BAS with IoT devices and smart thermostats can offer sophisticated control and optimization of the building's environment, ensuring comfort while minimizing energy use. Deploying innovations like VRF systems and advanced heat pumps can provide superior heating and cooling performance, adapting to the unique requirements of each space within the building. All these special considerations and advanced techniques enable HVAC contractors to provide cutting-edge, energy-efficient, and highly functional solutions to meet the diverse needs of modern buildings.

Below are concise overviews for each state in alphabetical order. Please remember to verify with the respective state's contractor licensing board or department for the most accurate and up-to-date information:

Alabama
- Requires license for projects over $50,000. Examination, financial statement, and proof of experience needed.

Alaska

- General contractors must be licensed. Requires bonds, insurance, and may require an exam.

Arizona
- Requires contractors to be licensed. Has strict examination and experience requirements.

Arkansas
- Requires license for projects over $20,000. Mandates examination, experience, and financial statement.

California
- Strict licensing requirements including examination, four years of experience, and fingerprinting.

Colorado
- Licensing is localized, with Denver having strict requirements including examinations and experience.

Connecticut
- Requires license for most contractors. Mandates examination and insurance.

Delaware
- Requires license, examination, and insurance. Registration with Division of Revenue also needed.

Florida
- Has stringent requirements including state examination and proof of financial stability.

Georgia
- Requires license for general and specialty contractors. Mandates examination and insurance.

Hawaii
- Requires license, examination, experience, and financial statements.

Idaho
- Lacks statewide licensing requirements for general contractors; localized regulations may apply.

Illinois
- Lacks statewide licensing for general contractors; localized regulations and examinations may apply.

Indiana
- Lacks statewide licensing for general contractors; some local jurisdictions have their own requirements.

Iowa
- Requires registration, insurance, and bonds. No statewide examination.

Kansas
- No statewide licensing for general contractors; localized requirements may exist.

Kentucky
- No statewide licensing for general contractors; some specialized contractors require licensing.

Louisiana
- Requires license for projects over $50,000. Mandates examination and financial statement.

Maine
- Lacks statewide licensing requirements; localized regulations may apply.

Maryland
- Requires home improvement contractors to be licensed, involving examination and financial stability proof.

Massachusetts
- Requires Construction Supervisor License, involving examinations and experience.

Michigan
- Requires Residential Builder or Maintenance & Alteration Contractor License, involving examination and insurance.

Minnesota
- Requires license, examination, insurance, and continuing education.

Mississippi
- Requires license for projects over $50,000; mandates examination and financial statement.

Missouri
- No statewide licensing for general contractors; localized regulations may exist.

Montana
- Requires registration; some specialized contractors require a license, involving examination and insurance.

Nebraska
- Requires registration; insurance and bond are mandatory.

Nevada
- Requires license, examination, experience, financial statement, and bond.

New Hampshire
- No statewide licensing for general contractors; some specialized contractors require licensing.

New Jersey
- Requires home improvement contractors to register; new home builders need a warranty.

New Mexico
- Requires license, examination, insurance, and bonding.

New York
- No statewide licensing for general contractors; localized regulations and registrations may apply.

North Carolina
- Requires license for projects over $30,000; mandates examination and financial statement.

North Dakota

- Requires license for projects over $4,000; mandates bond.

Ohio
- No statewide licensing for general contractors; some specialized contractors require licensing.

Oklahoma
- No statewide licensing for general contractors; some local jurisdictions have their own requirements.

Oregon
- Requires license, examination, insurance, and bond.

Pennsylvania
- Requires registration; insurance is mandatory.

Rhode Island
- Requires registration, insurance, and bond.

South Carolina
- Requires license, examination, and financial statement for projects over $5,000.

South Dakota
- No statewide licensing for general contractors; localized requirements may exist.

Tennessee
- Requires license for projects over $25,000; mandates examination and financial statement.

Texas
- No statewide licensing for general contractors; some specialized contractors require licensing.

Utah
- Requires license, examination, experience, insurance, and financial statement.

Vermont
- No statewide licensing for general contractors; some specialized contractors require licensing.

Virginia
- Requires license, examination, insurance, and financial statement.

Washington
- Requires registration, insurance, and bond.

West Virginia
- Requires license, examination, and financial statement for projects over $2,500.

Wisconsin
- Requires Dwelling Contractor Certification and license involving examination and insurance.

Wyoming
- No statewide licensing for general contractors; localized requirements may exist.

Even if a state lacks statewide licensing requirements, local municipalities or counties within that state may have their own licensing regulations, so always check locally. Additionally, specialty trades like plumbing, electrical, and HVAC almost always have strict licensing requirements, regardless of the regulations for general contractors.

Exam Prep Section:

Welcome to the Practice Test Question Section! This section is meticulously designed to give you a taste of the types of questions you'll encounter on the actual Contractor License exam. Our aim is to offer you a robust preparation tool that will test your knowledge, assess your understanding of the concepts, and enhance your ability to recall the information you've learned.

In this practice test section, we've chosen to place the correct answer and detailed explanation immediately following each question. We believe immediate feedback is crucial for learning and retention. By providing the answers and rationales right away, we aim to reinforce the correct information in your mind immediately, allowing you to understand not just the 'what', but also the 'why'. This approach helps to solidify your knowledge and clarify any misunderstandings on the spot, preventing the consolidation of incorrect information.

Every question is tailored to mimic the format and style of the actual exam, allowing you to familiarize yourself with the exam's structure. The rationale behind each answer is comprehensive, ensuring you grasp the underlying concepts and can apply them in the exam effectively.

Remember, the key to succeeding in the Contractor License exam is not merely memorizing information, but truly understanding it. This way, even if you encounter questions that are framed differently, your understanding of the concepts will empower you to choose the correct answer.

Effective Study Techniques:
1. **Creating a Study Plan:**
   - **Break Down the Syllabus:** Divide the syllabus into manageable sections and allocate specific time slots for each, allowing more time for challenging topics.
   - **Set Clear Goals:** Define what you aim to achieve in each study session. Concrete, achievable goals help in maintaining focus.
   - **Prioritize & Sequence:** Arrange topics in order of importance and difficulty, tackling the most challenging and crucial ones first.
   - **Regular Review:** Schedule time for revising previously studied materials to reinforce learning and aid retention.

2. **Active Learning Strategies:**
   - **Self-Testing:** Regularly challenge your understanding with practice tests and quizzes to identify areas needing improvement.
   - **Study Groups:** Collaborating with peers can provide different perspectives and can help in filling knowledge gaps.
   - **Mind Mapping:** Visual representations of information can help in understanding complex concepts and seeing the relationships between them.
   - **Teach Back:** Teaching learned concepts to someone else can reinforce understanding and uncover areas you may not have fully grasped.

Test-Taking Strategies:
1. **Time Management:**

- **Pace Yourself:** Allocate your time wisely, ensuring you have enough time to answer each question and review your work.
- **Prioritize:** Address easier questions first to secure those points, then move on to the more challenging ones.
- **Stay Aware of Time:** Regularly check the time and adjust your pace as needed to avoid rushing at the end.

2. **Handling Multiple Choice Questions:**
- **Read Carefully:** Fully understand the question and each answer choice before selecting an answer.
- **Elimination:** Rule out the obviously incorrect answers first, then make an educated guess among the remaining options if unsure.
- **Review:** If time permits, revisit questions to ensure you didn't make any mistakes or misread any questions.

3. **Reducing Test Anxiety:**
- **Regular Relaxation Techniques:** Incorporate deep breathing, progressive muscle relaxation, or meditation into your study routine to manage stress levels.
- **Positive Visualization:** Imagine a successful test-taking experience to build confidence and reduce negative thoughts.
- **Preparation:** Adequate preparation and practice can significantly reduce anxiety as familiarity and confidence in the material increase.
- **Healthy Lifestyle Choices:** Regular exercise, a balanced diet, adequate sleep, and staying hydrated are crucial for maintaining focus and reducing anxiety.

Review any specific test instructions, formats, and regulations provided by the exam authority in your preparations, and consider seeking additional resources or support if you struggle in any areas.

Now, let's dive into the practice questions! Good luck, and may each question bring you one step closer to acing your Contractor License exam!

1. A contractor is preparing a bid and needs to factor in overhead costs. Which of the following is not typically considered an overhead cost?
a. Employee Salaries
b. Rent for Office Space
c. Job Site Utilities
d. Insurance Premiums

Answer: c. Job Site Utilities. Explanation: Overhead costs refer to the ongoing costs to operate a business but are not directly tied to a specific construction project. Job Site Utilities are usually considered a direct cost associated with a specific project and would be factored into the overall project cost, not overhead.

2. A construction company is experiencing a cash flow crunch and needs to prioritize payments. Which payment should be addressed first in most scenarios?
a. Subcontractor Payments
b. Employee Wages
c. Office Rent
d. Equipment Lease

Answer: b. Employee Wages. Explanation: Failure to pay employee wages can lead to immediate legal consequences and can demoralize the workforce leading to productivity issues. It's crucial to maintain a good relationship with employees by ensuring their salaries are paid on time.

3. When analyzing a company's profit margin in construction, which of the following is an essential component to consider?
a. Gross Profit
b. Net Profit
c. Operating Profit
d. All of the Above

Answer: d. All of the Above. Explanation: Gross profit, net profit, and operating profit are all crucial components when analyzing a company's profit margin. They provide insight into different levels of profitability by accounting for various costs and expenses incurred by the business.

4. A construction project manager needs to create a risk management plan. Which step should be prioritized first?
a. Risk Identification
b. Risk Analysis
c. Risk Mitigation
d. Risk Monitoring

Answer: a. Risk Identification. Explanation: Identifying potential risks is the foundational step in risk management. It involves recognizing possible threats and uncertainties before assessing their impact, probability, and developing strategies to manage them.

5. While preparing a construction contract, which clause is vital to outline how potential disputes will be resolved?
a. Arbitration Clause
b. Indemnification Clause
c. Payment Clause
d. Scope of Work

Answer: a. Arbitration Clause. Explanation: An arbitration clause is crucial as it outlines the agreed-upon process for resolving disputes, which can help in avoiding lengthy and expensive court battles and ensure smoother resolution of conflicts.

6. When analyzing the critical path in a construction project schedule, which activity would be considered a "critical activity"?
a. An activity with float
b. An activity with no successors
c. An activity with no float
d. An activity with multiple successors

Answer: c. An activity with no float. Explanation: A critical activity is one that has zero float, meaning any delay in its completion will directly impact the project's finish date, making it essential to monitor and manage these activities closely.

7. A construction manager is faced with a situation where the project is significantly behind schedule. What should be the manager's first course of action?
a. Accelerate the project by increasing manpower
b. Inform the stakeholders of the delay
c. Analyze the reason for the delay
d. Request an extension from the client

Answer: c. Analyze the reason for the delay. Explanation: Understanding the root cause of the delay is critical before deciding on a course of action or communicating with stakeholders. Once the reason is identified, an effective solution can be devised and communicated to the relevant parties.

8. In construction project management, a cost variance (CV) that is less than zero typically indicates that the:
a. Project is under budget
b. Project is ahead of schedule
c. Project is over budget
d. Project cost is accurate

Answer: c. Project is over budget. Explanation: Cost Variance is calculated by subtracting the actual cost from the earned value (CV = EV - AC). A negative cost variance indicates that the actual cost is more than what was planned, signaling that the project is over budget.

9. A contractor is considering implementing lean construction principles. Which of the following is a primary focus of lean construction?
a. Cost Reduction
b. Waste Elimination
c. Schedule Acceleration
d. Quality Improvement

Answer: b. Waste Elimination. Explanation: Lean construction primarily focuses on eliminating waste in all forms, including time, materials, and labor, to enhance overall project value and efficiency.

10. A subcontractor has inadvertently breached a contract. What should the general contractor do first?
a. Terminate the contract
b. Seek legal counsel
c. Communicate with the subcontractor to understand the reasons
d. Claim damages immediately

Answer: c. Communicate with the subcontractor to understand the reasons. Explanation: Before taking any formal action, it is crucial to communicate with the subcontractor to understand the reasons behind the breach and explore possible remedies or resolutions amicably.

11. In a construction company looking to expand, which business structure is most suitable to attract investors due to its ability to issue stocks?
a. Sole Proprietorship
b. Limited Liability Company (LLC)
c. Corporation
d. Partnership

Answer: c. Corporation. Explanation: A corporation is the most suitable structure for attracting investors as it can issue stocks. It is a legal entity separate from its owners, providing them with liability protection and enabling easier access to capital.

12. When a construction company, structured as a Limited Liability Company (LLC), fails to keep personal and business finances separate, what risk does it run?
a. Increased Tax Liability
b. Losing Limited Liability Protection
c. Incurring Additional Operating Costs
d. Decreasing Profit Margins

Answer: b. Losing Limited Liability Protection. Explanation: Commingling of personal and business finances can lead to a piercing of the corporate veil, meaning the owners can lose their limited liability protection and become personally responsible for the business's debts.

13. A construction firm is looking to form a partnership. Which of the following agreements is crucial to outline the responsibilities and profit-sharing ratio among the partners?
a. Operating Agreement
b. Partnership Agreement
c. Articles of Organization
d. Bylaws

Answer: b. Partnership Agreement. Explanation: A Partnership Agreement is fundamental in a partnership structure. It outlines each partner's responsibilities, profit-sharing ratio, dispute resolution, and other essential aspects of the partnership.

14. Which business structure generally has the simplest setup and gives the owner full control but also entails personal liability for the company's debts?
a. Corporation
b. Limited Liability Company (LLC)
c. Sole Proprietorship
d. S Corporation

Answer: c. Sole Proprietorship. Explanation: Sole Proprietorship is the simplest business structure, allowing the owner full control. However, it does not offer liability protection, meaning the owner is personally responsible for the debts and liabilities of the business.

15. A contractor is working under a joint venture agreement. What is typically the main reason for choosing this business arrangement for a project?
a. Share Profits
b. Pool Resources and Expertise
c. Limit Liability
d. Expand Market Presence

Answer: b. Pool Resources and Expertise. Explanation: Joint ventures are usually formed to pool resources and expertise from different entities to undertake a specific project, allowing them to share risks, costs, and rewards.

16. When considering tax implications for a construction company, which business structure allows income to 'pass-through' to the owners' individual tax returns?
a. C Corporation
b. S Corporation
c. Limited Partnership
d. B Corporation

Answer: b. S Corporation. Explanation: An S Corporation is a pass-through entity, meaning the income, deductions, and credits of the corporation pass through to shareholders, who report this information on their individual tax returns.

17. A construction company is considering changing its business structure to a Limited Liability Partnership (LLP). Which of the following is a primary advantage of an LLP?
a. No Personal Liability for Business Debts
b. Ability to Issue Stocks
c. Exemption from Corporate Tax
d. Unlimited Number of Partners

Answer: a. No Personal Liability for Business Debts. Explanation: LLP provides protection to each partner from personal liability for business debts and the negligent acts of other partners, thus limiting their liability to their investment in the business.

18. A construction contractor operating as a sole proprietor is contemplating incorporating the business. What is a primary consideration regarding taxation in making this decision?
a. Double Taxation
b. Lower Tax Rates
c. Increased Tax Deductions
d. Avoidance of Self-Employment Tax

Answer: a. Double Taxation. Explanation: Corporations face the issue of double taxation, where the profits are taxed at the corporate level, and any distributed dividends are taxed again at the individual shareholder level.

19. A contractor forming a new construction company wants to retain management control while also protecting personal assets from business liabilities. Which business structure best meets this objective?
a. Sole Proprietorship
b. Corporation
c. Limited Liability Company (LLC)
d. General Partnership

Answer: c. Limited Liability Company (LLC). Explanation: An LLC combines the limited liability protection of a corporation with the management flexibility of a partnership, allowing the owner to protect personal assets while retaining control over business decisions.

20. In a construction company operating as an S Corporation, which of the following is a requirement related to shareholders?
a. Shareholders must be U.S. Citizens or Resident Aliens.
b. There can be unlimited shareholders.
c. Shareholders can be other corporations or LLCs.
d. There is no limit on the classes of stock that can be issued.

Answer: a. Shareholders must be U.S. Citizens or Resident Aliens. Explanation: For an S Corporation, shareholders must be U.S. Citizens or Resident Aliens, and there are restrictions on the number and type of allowable shareholders and the classes of stock that can be issued.

21. In a scenario where a contractor must allocate overhead costs to multiple construction projects, which method is most commonly used to ensure accuracy and fairness?
a. Direct Method
b. Step-Down Method
c. Reciprocal Allocation Method
d. Equal Allocation Method

Answer: c. Reciprocal Allocation Method. Explanation: The Reciprocal Allocation Method is used when different projects share overhead costs, ensuring that each project is allocated a fair and accurate portion of the costs, considering interdepartmental services.

22. When a construction company encounters an unexpected cash flow deficit, which financial strategy is typically most viable to manage short-term obligations?
a. Long-term Loan
b. Equity Financing
c. Line of Credit
d. Selling Fixed Assets

Answer: c. Line of Credit. Explanation: A Line of Credit is typically the most flexible and immediate solution for managing short-term cash flow needs, allowing companies to borrow only what they need when they need it.

23. How does underbilling impact a construction company's financial statements when using the percentage of completion method?
a. Decreases Assets and Increases Liabilities
b. Increases Assets and Decreases Equity
c. Increases Assets and Increases Liabilities
d. Decreases Assets and Decreases Equity

Answer: c. Increases Assets and Increases Liabilities. Explanation: Underbilling (billing less than the earned revenue) increases both assets (Underbillings) and liabilities (Billings in Excess of Costs) on the financial statement.

24. A construction contractor is reviewing the Profit and Loss Statement. Which item is not typically found on this statement?
a. Gross Profit
b. Operating Expenses
c. Net Income
d. Accounts Receivable

Answer: d. Accounts Receivable. Explanation: Accounts Receivable is found on the Balance Sheet, not the Profit and Loss Statement, which typically includes revenues, costs, expenses, and net income or loss.

25. When assessing the financial health of a construction company, what does a high Debt to Equity Ratio typically indicate?
a. Low Risk
b. High Leverage
c. Strong Liquidity
d. High Profitability

Answer: b. High Leverage. Explanation: A high Debt to Equity Ratio indicates that a company is using more debt to finance its operations, signifying high leverage and potentially higher risk.

26. In the construction industry, which financial ratio is crucial for assessing a company's ability to meet its short-term obligations?
a. Return on Equity
b. Current Ratio
c. Gross Margin Ratio
d. Debt to Equity Ratio

Answer: b. Current Ratio. Explanation: The Current Ratio measures a company's ability to pay off its short-term liabilities with its short-term assets, crucial for assessing liquidity in the construction industry.

27. How does the accrual basis of accounting impact a construction company's recognition of expenses and revenues?
a. Recognizes when cash is received or paid
b. Recognizes when earned or incurred
c. Recognizes when billed or invoiced
d. Recognizes when contracted or agreed upon

Answer: b. Recognizes when earned or incurred. Explanation: Accrual accounting recognizes revenues and expenses when they are earned or incurred, regardless of when cash transactions occur.

28. Which factor is vital for a construction company to consider when calculating the break-even point for a project?
a. Profit Margin
b. Fixed Costs
c. Variable Costs
d. All of the Above

Answer: d. All of the Above. Explanation: The break-even point is where total costs equal total revenues. Calculating it requires considering fixed costs, variable costs, and the profit margin on each unit.

29. In construction project management, what is the primary purpose of preparing a cash flow forecast?
a. To determine project profitability
b. To manage the timing of cash inflows and outflows
c. To calculate the net present value of the project
d. To estimate the total project cost

Answer: b. To manage the timing of cash inflows and outflows. Explanation: Cash flow forecasts are crucial for managing the timing of cash inflows and outflows, ensuring that the company can meet its financial obligations throughout the project.

30. When a construction company analyzes a project's Return on Investment (ROI), which of the following is a primary consideration?
a. Project Duration
b. Overhead Costs
c. Profit Earned relative to Investment Made
d. Variable Costs Incurred

Answer: c. Profit Earned relative to Investment Made. Explanation: ROI is a measure of the profitability of a project and is calculated as the profit earned relative to the investment made, helping to determine the financial viability of undertaking the project.

31. In construction contracts, what is the primary purpose of a 'time is of the essence' clause?
a. To mandate the completion of the project within a reasonable time frame
b. To stipulate that delays will not incur any penalties
c. To prioritize quality over speed in project completion
d. To define the scope of work in relation to the project timeline

Answer: a. To mandate the completion of the project within a reasonable time frame. Explanation: A 'time is of the essence' clause stresses the critical nature of meeting contractual deadlines, ensuring timely project completion.

32. When a contractor is required to make 'reasonable adjustments' for unforeseen issues, which contract type is likely in effect?
a. Lump Sum Contract
b. Cost Plus a Fee Contract
c. Unit Price Contract
d. Time and Material Contract

Answer: b. Cost Plus a Fee Contract. Explanation: Cost Plus a Fee contracts usually require contractors to make reasonable adjustments for unforeseen issues, as payment is determined by the actual cost of work plus a negotiated fee.

33. In a Unit Price Contract, what determines the final contract value?
a. Fixed price agreed upon in the contract
b. The actual quantity of work units completed
c. The estimated cost of materials and labor
d. The contractor's overhead and profit margin

Answer: b. The actual quantity of work units completed. Explanation: In a Unit Price Contract, the final contract value is determined by the actual number of work units completed multiplied by the unit price.

34. Which contract type is most suitable when project specifications and scope are not well-defined at the outset?
a. Lump Sum Contract
b. Fixed Price Contract
c. Time and Material Contract
d. Unit Price Contract

Answer: c. Time and Material Contract. Explanation: When scope and specifications are unclear, Time and Material contracts are suitable as they allow flexibility in adjusting labor hours and materials as the project evolves.

35. How does a Liquidated Damages clause typically impact a construction contract?
a. Increases project duration
b. Pre-determines the compensation for specific breaches, such as delays
c. Reduces contractor liability
d. Expands the scope of work

Answer: b. Pre-determines the compensation for specific breaches, such as delays. Explanation: Liquidated Damages clauses set a pre-agreed amount of compensation for specific breaches, providing a measure of financial relief and certainty for the injured party.

36. In managing construction contracts, which document is crucial for detailing the rights, responsibilities, and work agreements between the contractor and the subcontractor?
a. Notice to Proceed
b. Subcontractor Agreement
c. Change Order
d. Certificate of Substantial Completion

Answer: b. Subcontractor Agreement. Explanation: A Subcontractor Agreement clearly outlines the obligations, rights, and work arrangements between the main contractor and the subcontractor, protecting the interests of both parties.

37. Which document is essential for formalizing any alterations in the original contract terms, scope, or price?
a. Subcontractor Agreement
b. Change Order
c. Performance Bond
d. Warranty Deed

Answer: b. Change Order. Explanation: Change Orders are pivotal for officially documenting any modifications in the contractual scope, price, or terms, ensuring mutual agreement on the alterations made.

38. When managing contracts, why is it crucial for contractors to meticulously document project delays?
a. To facilitate effective project scheduling
b. To validate claims for extension of time or additional compensation
c. To optimize resource allocation and utilization
d. To monitor subcontractor performance and compliance

Answer: b. To validate claims for extension of time or additional compensation. Explanation: Precise documentation of delays is critical in substantiating requests for time extensions or extra payment due to disruptions or alterations in the project timeline.

39. In a construction project, how does the application of a Retention Holdback benefit the project owner?
a. It reduces the project cost and duration
b. It ensures the contractor's adherence to project quality and completion
c. It expands the project's scope of work
d. It accelerates the project completion timeline

Answer: b. It ensures the contractor's adherence to project quality and completion. Explanation: A Retention Holdback withholds a portion of the payment until project completion to ensure that the contractor fulfills all contractual obligations related to quality and completion.

40. In the context of contract management, what is the primary implication of a 'Pay if Paid' clause in a subcontractor agreement?
a. The subcontractor assumes the risk of non-payment by the project owner
b. The subcontractor receives payment directly from the project owner
c. The main contractor is obligated to pay the subcontractor, regardless of owner payment
d. The payment to the subcontractor is proportional to the work completed

Answer: a. The subcontractor assumes the risk of non-payment by the project owner. Explanation: A 'Pay if Paid' clause stipulates that the subcontractor will only be paid if the main contractor receives payment from the project owner, transferring the risk of non-payment to the subcontractor.

41. A construction manager, realizing the importance of morale in employee productivity, wants to address the primary needs of his team members as described by Maslow's Hierarchy of Needs. What should be his immediate focus for an employee who has job security and a fair salary?
a. Enhance team bonding and belongingness
b. Offer opportunities for professional growth
c. Secure a safer working environment
d. Provide more financial incentives

Answer: a. Enhance team bonding and belongingness. Explanation: Once basic and safety needs, such as salary and job security, are met, addressing social needs like belongingness and relationships is the next level in Maslow's Hierarchy of Needs.

42. When the supervisor of a construction project uses a transformational leadership style, what is likely to be the primary impact on the team?
a. Employees may feel overly managed due to constant supervision
b. The team is likely to experience increased motivation and morale
c. Workers may prefer a structured, directive approach
d. The team might feel neglected due to a lack of guidance

Answer: b. The team is likely to experience increased motivation and morale. Explanation: Transformational leadership inspires and motivates employees, encouraging innovation, and fostering a positive team environment, which usually leads to higher morale and motivation.

43. In the context of employee management, what is the principal objective of conducting performance appraisals for construction workers?
a. To determine salary increments
b. To allocate job responsibilities
c. To identify areas for improvement and development
d. To assign new projects

Answer: c. To identify areas for improvement and development. Explanation: Performance appraisals primarily aim to assess employee performance, provide feedback, and identify areas for improvement and development opportunities, although they may also influence salary and job assignments.

44. A construction manager decides to implement an employee recognition program aiming to increase productivity and job satisfaction. Which type of motivation theory is he applying?
a. Expectancy Theory
b. Herzberg's Two-Factor Theory
c. Maslow's Hierarchy of Needs
d. Equity Theory

Answer: b. Herzberg's Two-Factor Theory. Explanation: Herzberg's Two-Factor Theory emphasizes the role of motivator factors like recognition and achievement in enhancing job satisfaction and motivation, leading to increased productivity.

45. How can a construction manager best support employees' professional development and career progression within the company?
a. By enforcing strict compliance to rules
b. By offering opportunities for learning and development
c. By implementing a stringent performance appraisal system
d. By focusing solely on financial rewards and bonuses

Answer: b. By offering opportunities for learning and development. Explanation: Providing learning and development opportunities is crucial for supporting employees' professional growth, skill enhancement, and career progression within the organization.

46. In employee management, why is it critical to have clear and measurable objectives during the delegation of tasks?
a. To ensure fair distribution of workload
b. To facilitate easier employee termination
c. To establish clear expectations and accountability
d. To maintain a strict hierarchical structure

Answer: c. To establish clear expectations and accountability. Explanation: Clear and measurable objectives are vital for setting expectations, enabling employees to understand their responsibilities, and holding them accountable for their tasks.

47. When managing a diverse workforce in construction, what is a key consideration to foster inclusivity and a sense of belonging among employees?
a. Emphasizing uniformity and conformity
b. Offering identical training to all employees
c. Acknowledging and valuing individual differences
d. Enforcing strict hierarchical communication channels

Answer: c. Acknowledging and valuing individual differences. Explanation: Recognizing and respecting individual differences and unique perspectives is crucial for creating an inclusive environment and fostering a sense of belonging among diverse employees.

48. A construction company has implemented flexible working hours to enhance employee work-life balance. This approach primarily targets which category of needs in Maslow's Hierarchy?
a. Safety Needs
b. Esteem Needs
c. Physiological Needs
d. Social Needs

Answer: a. Safety Needs. Explanation: Offering flexible working hours contributes to meeting safety needs by reducing stress and enhancing the psychological well-being and work-life balance of employees.

49. In conflict resolution within a construction team, what is the primary advantage of using a collaborative approach?
a. It saves time by quickly resolving disputes
b. It ensures the manager's perspective is prioritized
c. It fosters mutual respect and understanding among team members
d. It maintains a strict and authoritative managerial stance

Answer: c. It fosters mutual respect and understanding among team members. Explanation: A collaborative approach to conflict resolution seeks to address the concerns of all parties involved, promoting mutual respect, understanding, and a harmonious working relationship.

50. In construction management, how can a leader effectively foster a positive and cohesive team environment?
a. By strictly enforcing company policies and procedures
b. By minimizing communication to avoid misunderstandings
c. By recognizing individual achievements and encouraging collaboration
d. By adopting a rigid and directive leadership style

Answer: c. By recognizing individual achievements and encouraging collaboration. Explanation: Acknowledging individual accomplishments and promoting teamwork and cooperation are pivotal in building a positive and cohesive team atmosphere.

51. In a scenario where a construction firm is exposed to various risks during a large-scale project, the firm opts to transfer some of the risks to a reputable insurance company. What is the most comprehensive type of policy that the firm should consider to cover multiple risks?
a. General Liability Policy
b. Builders Risk Insurance
c. Commercial Package Policy
d. Workers' Compensation Insurance

Answer: c. Commercial Package Policy. Explanation: Commercial Package Policy allows businesses to bundle various coverages, providing a comprehensive solution to manage multiple risks, tailored to the specific needs of the firm.

52. A construction manager decides to proceed with a project after carefully considering the potential negative outcomes and benefits. What risk management strategy is being employed here?
a. Risk avoidance
b. Risk transfer
c. Risk retention
d. Risk acceptance

Answer: d. Risk acceptance. Explanation: When a manager decides to proceed with a project, understanding the potential negative outcomes and deeming the benefits worthwhile, it is known as risk acceptance.

53. In the case where a subcontractor fails to complete their portion of work on time, leading to delays in project completion, which type of insurance can protect the general contractor from claims related to delay damages?
a. Professional Liability Insurance
b. Performance Bond
c. Builders Risk Insurance
d. Employment Practices Liability Insurance

Answer: b. Performance Bond. Explanation: Performance Bonds assure the contractor's performance according to the contract terms, including completion times, and protect against losses due to delays or failure to perform.

54. How can construction companies primarily minimize the risk of financial loss due to employees' injuries on the job?
a. By implementing rigorous employee training programs
b. By purchasing Workers' Compensation Insurance
c. By enforcing strict project deadlines
d. By conducting regular financial audits

Answer: b. By purchasing Workers' Compensation Insurance. Explanation: Workers' Compensation Insurance is specifically designed to cover medical expenses and lost wages due to work-related injuries, thereby minimizing financial loss for companies.

55. When developing a risk management plan, what should be the primary focus of a construction manager regarding identified risks?
a. Allocation of resources to eliminate all risks
b. Prioritizing risks based on their impact and probability
c. Transferring all risks to subcontractors
d. Ignoring minor risks to focus solely on major ones

Answer: b. Prioritizing risks based on their impact and probability. Explanation: Risks should be prioritized based on their potential impact and likelihood, allowing the allocation of resources to manage and mitigate the most significant risks effectively.

56. A construction firm is concerned about potential claims of damage to third-party property during a project. What insurance policy should the firm secure to mitigate this risk?
a. Employment Practices Liability Insurance
b. Professional Liability Insurance
c. General Liability Insurance
d. Builders Risk Insurance

Answer: c. General Liability Insurance. Explanation: General Liability Insurance protects against claims of bodily injury and property damage to third parties due to the firm's operations, hence mitigating the risk of such claims.

57. In a high-risk project, a contractor opts to share the risk with another construction firm. What risk management strategy is the contractor employing?
a. Risk Avoidance
b. Risk Transfer
c. Risk Sharing
d. Risk Retention

Answer: c. Risk Sharing. Explanation: Risk sharing involves distributing the risks between multiple parties, such as partnering with another firm, to minimize the impact of potential negative outcomes on any single entity.

58. For a construction project manager, why is it essential to conduct a comprehensive risk assessment at the initial stages of a project?
a. To allocate all risks to insurance companies
b. To identify and analyze potential risks and develop mitigation strategies
c. To determine the profitability of the project
d. To decide on the project deadlines

Answer: b. To identify and analyze potential risks and develop mitigation strategies. Explanation: Conducting a comprehensive risk assessment early allows the identification and analysis of potential risks, enabling the development of effective mitigation and management strategies.

59. When a construction company employs stringent safety protocols and regular training programs to minimize the risk of on-site accidents, which risk management technique is being utilized?
a. Risk Transfer
b. Risk Avoidance
c. Risk Mitigation
d. Risk Acceptance

Answer: c. Risk Mitigation. Explanation: By implementing safety measures and training, the company is actively working to reduce the likelihood and/or impact of on-site accidents, thereby mitigating risk.

60. In dealing with potential legal disputes and litigation risks in construction projects, which contract clause is crucial for defining the resolution mechanism between involved parties?
a. Indemnity clause
b. Payment clause
c. Dispute resolution clause
d. Scope of work clause

Answer: c. Dispute resolution clause. Explanation: The dispute resolution clause in a contract specifies the mechanism for resolving disagreements between the parties, such as arbitration or litigation, providing a predefined pathway to address potential legal disputes.

61. If a construction company recognizes income on an accrual basis for tax purposes, in what scenario is the income considered earned?
a. When the client signs the contract.
b. When the payment is received from the client.
c. When the invoice is sent to the client.
d. When the company completes the agreed-upon work.

Answer: d. When the company completes the agreed-upon work. Explanation: Under the accrual basis, income is recognized when it is earned, i.e., when the services are rendered or the product is delivered, not when payment is received.

62. A construction firm incorrectly classified several employees as independent contractors. What type of penalty may this misclassification result in?
a. Overpayment of income tax.
b. Liability for employment taxes.
c. Reduced payroll tax.
d. Ineligibility for tax deductions.

Answer: b. Liability for employment taxes. Explanation: Misclassifying employees as independent contractors can result in liability for employment taxes, including both the employer and employee portions of Social Security and Medicare taxes.

63. A construction manager decided to lease equipment for a new project due to limited capital. For tax purposes, what should the manager consider regarding the lease payments?
a. They are a non-deductible expense.
b. They are a capital expenditure.
c. They are deductible as a business expense.
d. They can be deducted as depreciation.

Answer: c. They are deductible as a business expense. Explanation: Lease payments for business equipment are generally considered a business expense and are deductible.

64. When a construction company pays its employees, what taxes must the employer withhold from the employees' paychecks?
a. Federal income tax, State income tax, and Property tax
b. Federal income tax, Social Security tax, and Medicare tax
c. State income tax, Property tax, and Sales tax
d. Sales tax, Federal income tax, and Social Security tax

Answer: b. Federal income tax, Social Security tax, and Medicare tax. Explanation: Employers are required to withhold Federal income tax, Social Security tax, and Medicare tax from employees' paychecks.

65. If a contractor is subject to a progressive tax system, how is their income taxed?
a. At a uniform rate.
b. At an increasing rate for higher income levels.
c. At a decreasing rate for higher income levels.
d. At the corporate tax rate.

Answer: b. At an increasing rate for higher income levels. Explanation: In a progressive tax system, the rate of tax increases as the amount of the taxpayer's income increases.

66. When a contractor makes improvements to a property which extends its useful life, how should the costs associated with the improvements be treated for tax purposes?
a. Expensed in the current tax year.
b. Deducted as a repair expense.
c. Capitalized and depreciated over time.
d. Deducted as an indirect cost.

Answer: c. Capitalized and depreciated over time. Explanation: Costs that extend the useful life of a property should be capitalized and depreciated over the useful life of the improvement.

67. Which statement is true regarding the payment of payroll taxes by a construction company?
a. They are solely the responsibility of the employee.
b. They are the responsibility of both the employer and the employee.
c. They are deductible as a business expense by the employee.
d. They are optional and can be deferred.

Answer: b. They are the responsibility of both the employer and the employee. Explanation: Payroll taxes are shared responsibilities, with both employer and employee contributing to Social Security and Medicare taxes.

68. A construction company is liable for an underpayment of estimated taxes. Which of the following is a consequence the company might face?
a. The company will receive a tax credit.
b. The company will face an interest penalty.
c. The company's tax bracket will be lowered.
d. The company will be exempt from paying taxes the following year.

Answer: b. The company will face an interest penalty. Explanation: If a company has underpaid estimated taxes, it will typically face an interest penalty until the underpaid amount is satisfied.

69. For a construction firm, which of the following expenditures would be considered a deductible business expense?
a. Personal vehicle expenses
b. Capital improvements
c. Purchase of investment property
d. Employee salaries

Answer: d. Employee salaries. Explanation: Employee salaries are considered a normal, necessary, and ordinary business expense and are thus deductible.

70. A contractor performing a renovation notes a significant discrepancy between the recorded and actual amount of materials on hand. What step should be taken first in resolving this discrepancy?
a. Adjust the financial statements to reflect the actual amount.
b. Investigate to determine the cause of the discrepancy.
c. Reorder materials to match the recorded amount.
d. Ignore the discrepancy as it is likely a minor error.

Answer: b. Investigate to determine the cause of the discrepancy. Explanation: Before making adjustments, it is crucial to investigate discrepancies to determine whether it is due to theft, loss, error in recording, or other reasons and address the root cause appropriately.

71. What documentation is typically required to demonstrate financial stability when applying for a contractor's license?
a. Recent pay stubs.
b. Personal and business bank statements.
c. A signed affidavit attesting to financial stability.
d. A letter of recommendation from a financial institution.

Answer: b. Personal and business bank statements.
Explanation: Personal and business bank statements are usually required to demonstrate financial stability and integrity during the licensing process, as they give a detailed view of income, expenses, and overall financial health.

72. In a state where continuing education is required for license renewal, what might happen if a contractor fails to meet the education requirement before the license expiration date?
a. The contractor can no longer bid on projects.
b. The contractor's license may not be renewed.
c. The contractor's business will be audited.
d. The contractor will receive a financial penalty, but the license will be renewed.

Answer: b. The contractor's license may not be renewed.
Explanation: In many states, if continuing education requirements are not met, the license will not be renewed, thus the contractor cannot legally continue to operate.

73. What is the likely consequence for a contractor working without a valid license?
a. A written warning.
b. A temporary license suspension.
c. Fines and potential legal action.
d. Mandatory continuing education.

Answer: c. Fines and potential legal action.
Explanation: Operating without a valid license is a serious violation and usually results in fines, legal actions, and potential prohibition from obtaining a license in the future.

74. When submitting a bid, a contractor discovers a requirement for a specific license endorsement he does not possess. What should the contractor do?
a. Submit the bid and obtain the endorsement if the bid is successful.
b. Obtain the endorsement before submitting the bid.
c. Submit the bid without the endorsement, disclosing this fact.
d. Partner with a contractor who has the required endorsement and submit a joint bid.

Answer: b. Obtain the endorsement before submitting the bid. Explanation: A contractor should always ensure they meet all the requirements and possess all necessary endorsements before submitting a bid to avoid legal complications and ensure compliance.

75. A contractor's license has expired. What is the typical protocol for regaining licensure?
a. The contractor must retake the licensure examination.
b. The contractor must pay a renewal fee and may face additional penalties.
c. The contractor must complete additional continuing education hours.
d. The contractor can no longer regain licensure.

Answer: b. The contractor must pay a renewal fee and may face additional penalties. Explanation: Typically, an expired license can be renewed by paying the necessary renewal fees and any late fees or penalties that may have been accrued, but processes can vary by jurisdiction.

76. A contractor is operating in a state with a reciprocity agreement with neighboring states. What does this mean for the contractor's licensing requirements?
a. The contractor can operate in any state without additional licensing.
b. The contractor can operate in neighboring states but with limited scope.
c. The contractor's license is valid in the states with the agreement, but additional documentation may be required.
d. The contractor needs to obtain a federal license to operate in multiple states.

Answer: c. The contractor's license is valid in the states with the agreement, but additional documentation may be required. Explanation: Reciprocity agreements usually mean the contractor can operate in the neighboring states with the agreement, but typically there will still be some form of application or documentation process to legally operate in those states.

77. A contractor is unsure if his business activities in another state require licensure in that state. What should he do to clarify this?
a. Consult the state's contractor licensing board or authority.
b. Ask other contractors operating in that state.
c. Continue operating until notified otherwise.
d. Obtain a license to be on the safe side without inquiring.

Answer: a. Consult the state's contractor licensing board or authority. Explanation: The contractor should consult directly with the state's licensing authority to obtain accurate information about licensing requirements.

78. A newly licensed contractor wishes to operate under a DBA (Doing Business As) name. What is typically required for this?
a. Submitting the DBA name when applying for the contractor's license.
b. Filing a separate application with the state contractor's board.
c. Registering the DBA with the appropriate local or state agency.
d. Obtaining approval from the IRS.

Answer: c. Registering the DBA with the appropriate local or state agency.
Explanation: A DBA name usually requires registration with the local or state agency responsible for business registrations, separate from the contractor licensing process.

79. A contractor is applying for a license in a state that requires proof of insurance. Which types of insurance are typically required?
a. Health and life insurance.
b. Auto and property insurance.
c. Liability and workers' compensation insurance.
d. Homeowner's and renter's insurance.

Answer: c. Liability and workers' compensation insurance.
Explanation: States typically require proof of liability and workers' compensation insurance to protect clients and employees.

80. In applying for a contractor's license, why might a contractor be required to submit a credit report?
a. To determine the contractor's bid limit.
b. To assess the contractor's financial responsibility and stability.
c. To evaluate the contractor's eligibility for state-sponsored projects.
d. To calculate the contractor's license renewal fees.

Answer: b. To assess the contractor's financial responsibility and stability.

81. When a project manager realizes that a project is behind schedule due to unforeseen site conditions, which of the following should be immediately performed?
a. Crash the schedule.
b. Request additional funds.
c. Update the risk management plan.
d. Communicate with stakeholders and update the project plan.

Answer: d. Communicate with stakeholders and update the project plan.
Explanation: Communication is crucial in project management. The project manager should immediately inform stakeholders of the situation, reassess, and adjust the project plan accordingly.

82. In construction scheduling, the Critical Path Method (CPM) is used to determine the:
a. Least important activities in the project.
b. Longest sequence of activities in the project.
c. Most resource-intensive activities in the project.
d. Shortest sequence of activities in the project.

Answer: b. Longest sequence of activities in the project.
Explanation: The CPM is utilized to identify the longest sequence of tasks in a project, which determines the shortest possible duration to complete the project.

83. When a contractor reviews a project's Earned Value Analysis and discovers a Cost Variance (CV) of - 10,000, this indicates that the project is:
a. Under budget by $10,000.
b. Over budget by $10,000.
c. Ahead of schedule by $10,000 worth of work.
d. Behind schedule by $10,000 worth of work.

Answer: b. Over budget by $10,000.
Explanation: A negative CV indicates that the project is over budget. The value depicts the amount by which actual costs exceed the planned budget.

84. In the case where multiple projects are running concurrently, which project management technique would be most effective to allocate resources?
a. PERT Analysis.
b. Critical Path Method.
c. Resource Leveling.
d. Monte Carlo Simulation.

Answer: c. Resource Leveling.
Explanation: Resource leveling is a technique in project management that involves adjusting the project schedule and allocating resources in a balanced manner, which is especially crucial when handling multiple projects.

85. In a scenario where a client requests additional work not included in the original contract, what should the contractor issue?
a. A new contract.
b. A work stoppage notice.
c. A change order.
d. An addendum.

Answer: c. A change order.
Explanation: A change order is issued when there are alterations to the scope of work, including additional requests from the client, that were not included in the original contract.

86. When utilizing the Last Planner System (LPS) in lean construction, the primary focus is on:
a. Cost Estimation.
b. Resource Allocation.
c. Work Planning and Control.
d. Risk Management.

Answer: c. Work Planning and Control.
Explanation: The Last Planner System is a scheduling system focused on creating reliable workflow and enhancing project performance, primarily focusing on planning and control of the work.

87. To handle uncertainties in project scheduling, which scheduling technique should be applied?
a. Critical Path Method.
b. Earned Value Management.
c. PERT (Program Evaluation Review Technique).
d. Gantt Chart.

Answer: c. PERT (Program Evaluation Review Technique).
Explanation: PERT is particularly useful for projects with a high degree of uncertainty, as it employs probabilistic time estimates to understand the variability in the project schedules.

88. When there are unexpected delays due to severe weather conditions, what type of delay is this considered to be in project management terms?
a. Excusable and compensable delay.
b. Non-excusable and non-compensable delay.
c. Excusable and non-compensable delay.
d. Non-excusable and compensable delay.

Answer: c. Excusable and non-compensable delay.
Explanation: Weather conditions are generally considered excusable since they are beyond the control of the contractor, but they are usually non-compensable as they are common risks in construction projects.

89. If a contractor plans to complete a project in phases due to site restrictions, which scheduling technique is best suited for depicting relationships between project activities and phases?
a. Gantt Chart.
b. Milestone Chart.
c. Network Diagram.
d. Histogram.

Answer: c. Network Diagram.
Explanation: Network Diagrams are effective in displaying the dependencies and relationships between project activities and are particularly useful when dealing with phased or segmented projects.

90. When a project manager realizes that the team is consistently completing tasks faster than scheduled, the schedule should be:
a. Left as is, as it allows for contingencies.
b. Adjusted to allocate additional resources to other projects.
c. Compressed to reduce the project timeline.
d. Extended to add additional tasks.

Answer: c. Compressed to reduce the project timeline.
Explanation: If the team is consistently ahead of schedule, the project manager might opt to compress the schedule, thus reducing the overall project timeline and delivering the project earlier than planned.

91. In a scenario where a contractor has accidentally damaged a neighboring property during construction, which of the following actions is most ethically and legally sound?
a. Repair the damage immediately without informing the property owner.
b. Inform the property owner and repair the damage immediately.
c. Ignore the damage as it was accidental.
d. Inform the property owner and wait for their response before taking any action.

Answer: b. Inform the property owner and repair the damage immediately.
Explanation: Taking immediate responsibility by informing the property owner and repairing the damage demonstrates ethical behavior and legal responsibility.

92. When it is discovered that a subcontractor is employing undocumented workers, the contractor should:
a. Ignore as it is the subcontractor's responsibility.
b. Report the subcontractor to the relevant authorities.
c. Terminate the contract with the subcontractor.
d. Request the subcontractor to rectify the situation immediately.

Answer: d. Request the subcontractor to rectify the situation immediately.
Explanation: The contractor should address this serious legal violation by giving the subcontractor the opportunity to rectify the situation, ensuring compliance with all relevant laws.

93. A contractor, to gain an unfair advantage in a bid, offers a gift to the client's procurement officer. This action is considered:
a. Acceptable if the gift is of minimal value.
b. Unethical and possibly illegal.
c. A common practice in competitive bidding.
d. Acceptable if the gift is reported.

Answer: b. Unethical and possibly illegal.
Explanation: Offering gifts to gain an unfair advantage is a breach of ethical conduct and could result in legal consequences, depending on the jurisdiction's bribery and corruption laws.

94. If a contractor is aware of a design flaw in the project but decides to remain silent, this decision can be categorized as:
a. Ethically sound, to avoid project delays.
b. A breach of contractual obligations.
c. Legally sound, as it is the designer's responsibility.
d. Acceptable, if it does not compromise safety.

Answer: b. A breach of contractual obligations.
Explanation: Contractors have a duty to inform all stakeholders of any known flaws or errors, not disclosing such information is both ethically and legally problematic.

95. When a competitor's bid is accidentally sent to a contractor, the ethical action would be to:
a. Review the bid for competitive analysis.
b. Inform the competitor and delete the bid.
c. Use the information to modify their bid.
d. Forward the bid to the client for transparency.

Answer: b. Inform the competitor and delete the bid.
Explanation: The ethical response is to inform the competitor about the mistake and delete the bid without reviewing it, maintaining integrity and fairness in the bidding process.

96. In case a client requests a contractor to cut corners to save costs, the contractor should:
a. Agree, to maintain good client relations.
b. Refuse and adhere to quality and safety standards.
c. Agree but secretly maintain quality standards.
d. Negotiate a middle ground for cost-saving with quality.

Answer: b. Refuse and adhere to quality and safety standards.
Explanation: Contractors should uphold quality and safety standards regardless of client requests to compromise them, ensuring legal compliance and ethical responsibility.

97. In construction contracts, the legal principle "Let the buyer beware" is referred to as:
a. Quantum meruit.
b. Exculpatory clause.
c. Caveat emptor.
d. Prima facie.

Answer: c. Caveat emptor.
Explanation: "Caveat emptor" means "let the buyer beware," emphasizing that the buyer must perform due diligence before making a purchase as they are buying at their own risk.

98. A contractor that overbills a client and justifies it as a common industry practice is exhibiting:
a. Ethical flexibility.
b. Legal business acumen.
c. Unethical and illegal behavior.
d. Standard risk mitigation.

Answer: c. Unethical and illegal behavior.
Explanation: Overbilling is both unethical and illegal, regardless of how common it may be perceived within the industry, and can lead to severe legal repercussions and damage to reputation.

99. When there's a disagreement between a contractor and a client, before proceeding to litigation, they should ideally explore:
a. An ultimatum.
b. Alternative dispute resolution methods.
c. Immediate contract termination.
d. Public arbitration.

Answer: b. Alternative dispute resolution methods.
Explanation: It is usually beneficial for both parties to explore alternative dispute resolution methods such as mediation or arbitration before resorting to litigation, which can be time-consuming and costly.

100. If a contractor notices that the worksite is not in compliance with OSHA standards, the contractor is ethically bound to:
a. Continue work but document the violations.
b. Report the violations to OSHA immediately.
c. Stop work and rectify the violations promptly.
d. Inform the client and request additional payment to rectify the violations.

Answer: c. Stop work and rectify the violations promptly.
Explanation: Compliance with OSHA standards is mandatory, and any violation should be corrected immediately to ensure the safety and well-being of all workers on the site.

101. A contractor is approached by a client to construct a building that does not conform to local zoning ordinances. What is the correct course of action?
a. Agree to construct, but require the client to handle any legal consequences.
b. Refuse the project and inform local authorities.
c. Construct the building but modify the design to adhere to local zoning ordinances.
d. Refuse the project and advise the client to seek a variance or re-zoning.

Answer: d. Refuse the project and advise the client to seek a variance or re-zoning.
Explanation: Contractors should operate within the law and advise clients to obtain the necessary permits and approvals before proceeding with non-conforming projects.

102. A contractor has been found to have violated environmental regulations during construction. What is the likely legal outcome?
a. License revocation and possible fines.
b. A warning and mandatory training.
c. Community service and reparations to the affected area.
d. Informal settlement with the environmental agency.

Answer: a. License revocation and possible fines.
Explanation: Violating environmental regulations can lead to serious consequences including license revocation and substantial fines, depending on the severity and impact of the violation.

103. In the event a contractor fails to pay subcontractors, what legal mechanism can subcontractors employ to secure payment?
a. Place a mechanic's lien on the property.
b. Seek an injunction against the contractor.
c. Apply for a restitution order.
d. Request arbitration with the contractor.

Answer: a. Place a mechanic's lien on the property.
Explanation: A mechanic's lien is a legal tool that subcontractors can use to secure payment by creating a security interest in the property until they are paid.

104. When a contractor knowingly uses substandard materials to cut costs, he is likely committing:
a. Breach of warranty.
b. Constructive fraud.
c. Negligent misrepresentation.
d. Strict liability.

Answer: b. Constructive fraud.
Explanation: Using substandard materials knowingly is considered constructive fraud, as it involves deceiving the client about the quality and compliance of the materials used.

105. In terms of contractual obligations, the legal doctrine "Time is of the Essence" implies:
a. The contract has a flexible timeline.
b. The contract must be completed in a reasonable time frame.
c. Specific deadlines in the contract are strictly binding.
d. Extensions of time are not permissible under the contract.

Answer: c. Specific deadlines in the contract are strictly binding.
Explanation: "Time is of the Essence" implies that the timelines specified in the contract are critical, and failure to adhere to them can be considered a material breach of contract.

106. If a contractor's employee is injured on the job due to lack of proper safety measures, which legislation primarily comes into play?
a. Fair Labor Standards Act (FLSA)
b. Occupational Safety and Health Act (OSHA)
c. Employee Retirement Income Security Act (ERISA)
d. Family and Medical Leave Act (FMLA)

Answer: b. Occupational Safety and Health Act (OSHA)
Explanation: OSHA primarily governs workplace safety, and a lack of proper safety measures that leads to an injury would generally be addressed under this act.

107. A client insists on a clause in the contract that relieves the contractor of liability for any negligent acts. Such a clause is generally considered:
a. Valid and enforceable.
b. Unconscionable and void.
c. Subject to arbitration.
d. Valid if both parties agree.

Answer: b. Unconscionable and void.
Explanation: Clauses that attempt to relieve a party of liability for their own negligence are typically considered unconscionable and unenforceable due to public policy considerations.

108. In case of a dispute over defective workmanship, the doctrine of "Substantial Performance" allows a contractor to:
a. Claim full payment less the cost to repair the defect.
b. Void the contract without penalties.
c. Demand arbitration to resolve the dispute.
d. Request additional time to rectify the defect.

Answer: a. Claim full payment less the cost to repair the defect.
Explanation: The doctrine of "Substantial Performance" acknowledges that, despite minor defects, if the main objective of the contract has been fulfilled, the contractor is entitled to payment, less any deductions for repairs.

109. When a contractor's license is expired, and they continue to operate, the legal consequence is typically:
a. A warning and a grace period to renew.
b. Immediate arrest and incarceration.
c. Fines and possible revocation of the license.
d. Mandatory completion of additional training.

Answer: c. Fines and possible revocation of the license.
Explanation: Operating with an expired license can result in substantial fines and potentially lead to permanent revocation of the license, depending on the jurisdiction's regulations and the specific circumstances.

110. If a contractor intentionally underbids a project and subsequently demands more money to complete it, he can be found guilty of:
a. Constructive eviction.
b. Bid rigging.
c. Quantum meruit.
d. Unjust enrichment.

Answer: b. Bid rigging.
Explanation: Intentionally underbidding to secure a project and then demanding more money could be considered a form of bid rigging, which is illegal and can lead to severe penalties.

111. When encountering a discrepancy between state and local building codes, a contractor should:
a. Follow the less restrictive code.
b. Prioritize the state code over local code.
c. Always follow the more restrictive code.
d. Seek legal counsel before proceeding.

Answer: c. Always follow the more restrictive code.
Explanation: Contractors are usually required to comply with both state and local laws, but when they conflict, the more restrictive law generally takes precedence to ensure the highest standard of safety and compliance.

112. A contractor working on a residential project in a historic district should be aware that:
a. Standard building codes apply without any modifications.
b. Only aesthetic modifications are allowed.
c. Specific restrictions and guidelines often apply to maintain historical integrity.
d. Prior approval from state agencies is not required.

Answer: c. Specific restrictions and guidelines often apply to maintain historical integrity.
Explanation: Historic districts often have specific guidelines and restrictions in place to preserve the architectural and historic integrity of the area. Contractors must adhere to these to avoid legal complications.

113. If a contractor knowingly violates local zoning ordinances, they could face:
a. Mandatory re-zoning applications.
b. Fines, project dismantlement, or both.
c. A lenient warning and a chance to rectify.
d. A mere citation without any financial penalty.

Answer: b. Fines, project dismantlement, or both.
Explanation: Knowingly violating zoning ordinances can result in severe penalties including fines and may also require the contractor to dismantle any non-compliant work.

114. Which is a common recourse when a contractor performs work without the necessary permits?
a. A retrospective approval process.
b. Fines and possible mandatory correction of the work.
c. Arbitration with local building authorities.
d. An informal warning and guidance for future compliance.

Answer: b. Fines and possible mandatory correction of the work.
Explanation: Performing work without the necessary permits typically leads to fines and may require the contractor to correct or remove any work that does not comply with local regulations.

115. In a region with strict environmental protection laws, a contractor must prioritize:
a. Cost-effective construction methods.
b. Rapid project completion.
c. Compliance with environmental protection measures.
d. Utilization of locally sourced materials.

Answer: c. Compliance with environmental protection measures.
Explanation: In areas with strict environmental laws, adhering to these regulations is crucial to avoid legal complications, fines, and potential project disruptions, even if it impacts cost and timelines.

116. A contractor hired to renovate a public building must ensure that modifications comply with:
a. The original building design.
b. ADA (Americans with Disabilities Act) standards.
c. The preferences of the building's occupants.
d. The contractor's aesthetic judgment.

Answer: b. ADA (Americans with Disabilities Act) standards.
Explanation: Renovations to public buildings must comply with ADA standards to ensure accessibility and usability for individuals with disabilities.

117. A state requires adherence to seismic building codes. What is a contractor's primary responsibility?
a. Avoiding additional construction costs.
b. Obtaining seismic risk insurance.
c. Ensuring structural integrity to withstand seismic activity.
d. Seeking waivers for minor seismic compliance.

Answer: c. Ensuring structural integrity to withstand seismic activity.
Explanation: In areas with seismic building codes, the contractor's primary responsibility is to ensure that constructions are capable of withstanding seismic activity to protect life and property.

118. If a local ordinance mandates the use of fire-resistant materials in construction, failure to comply can result in:
a. A relaxation of standards.
b. Fines, demolition orders, or legal actions.
c. A compromise with local building authorities.
d. Approval with minor modifications.

Answer: b. Fines, demolition orders, or legal actions.
Explanation: Non-compliance with local ordinances regarding the use of specific materials can lead to serious consequences, including fines, orders for demolition, and legal repercussions.

119. When encountering local laws that mandate sustainable building practices, the contractor's primary concern should be:
a. Identifying least-cost sustainable methods.
b. Adherence to and implementation of sustainable building practices.
c. Negotiation for relaxation of sustainable mandates.
d. Rapid completion to avoid additional sustainable compliance.

Answer: b. Adherence to and implementation of sustainable building practices.
Explanation: Contractors must prioritize compliance with local laws mandating sustainable building practices to meet environmental standards and avoid legal implications.

120. A contractor working in a flood-prone area should primarily be concerned with:
a. Fast-tracking the construction process.
b. Securing flood insurance for the project.
c. Complying with flood-resistant construction standards.
d. Identifying alternate construction sites.

Answer: c. Complying with flood-resistant construction standards.
Explanation: In flood-prone areas, contractors must adhere to construction standards designed to resist flood damage to protect properties and inhabitants and avoid legal ramifications.

121. If a contractor and a homeowner enter into a contractual agreement, and the homeowner revokes it unilaterally, the contractor:
a. Has no recourse as the homeowner has the right to revoke.
b. Can only claim for the works that have been completed.
c. Can seek damages if there is a breach of contract.
d. Must return any advanced payment received immediately without any claim.

Answer: c. Can seek damages if there is a breach of contract.
Explanation: Contractors may seek damages if a unilateral revocation results in a breach of contract, depending on the terms outlined in the agreement.

122. When a contractor enters into a contract that includes an indemnity clause, the contractor is obligated to:
a. Complete the project within the specified timeframe.
b. Hold the indemnified party harmless against specified losses or damages.
c. Secure additional insurance coverage.
d. Obtain performance bonds for the project.

Answer: b. Hold the indemnified party harmless against specified losses or damages.
Explanation: Indemnity clauses usually require the indemnifying party to compensate the indemnified party for certain losses or damages as specified in the contract.

123. A contractor who agrees to a liquidated damages clause in a contract should understand that:
a. It precludes any other form of damages.
b. It establishes a pre-determined amount to be paid for each day of delay.
c. It eliminates the contractor's liability for delays.
d. It is non-binding and can be disputed in court.

Answer: b. It establishes a pre-determined amount to be paid for each day of delay.
Explanation: Liquidated damages clauses stipulate a predetermined amount of money that must be paid for each day the completion of the work is delayed beyond the contract completion date.

124. When discrepancies occur between contract drawings and specifications, the contractor should:
a. Prioritize the specifications over the drawings.
b. Resolve the discrepancy by seeking clarifications from the relevant authority.
c. Continue work as per their best judgment.
d. Use the most cost-effective interpretation.

Answer: b. Resolve the discrepancy by seeking clarifications from the relevant authority.
Explanation: Contractors should seek clarifications from architects, engineers, or other authoritative entities to resolve discrepancies between contract drawings and specifications to avoid future disputes and ensure compliance.

125. A contractor enters into a contract which states that any modifications to the contract must be in writing. However, the client requests an oral modification and promises to pay. The contractor should:
a. Proceed with the oral modification, relying on the client's promise.
b. Decline the oral modification and insist on a written one.
c. Charge an additional fee for the oral modification.
d. Complete the oral modification but withhold delivery until written confirmation is received.

Answer: b. Decline the oral modification and insist on a written one.
Explanation: To maintain contractual integrity and avoid disputes, the contractor should insist that any modifications to the contract be made in writing as per the original contract agreement.

126. A construction contract stipulates a "pay when paid" clause. This means that the contractor:
a. Must pay the subcontractors only when the contractor gets paid.
b. Is obligated to pay subcontractors irrespective of receiving payment.
c. Can delay payment indefinitely.
d. Has unilateral discretion on payment timings.

Answer: a. Must pay the subcontractors only when the contractor gets paid.
Explanation: A "pay when paid" clause typically means the general contractor is only obligated to pay the subcontractors when or if they receive payment from the owner.

127. A contractor fails to properly review a contract and misses a crucial detail which results in a financial loss. This demonstrates a lack of:
a. Ethical compliance.
b. Contractual capacity.
c. Due diligence.
d. Financial liquidity.

Answer: c. Due diligence.
Explanation: A failure to thoroughly review a contract and subsequently overlooking crucial details demonstrates a lack of due diligence on the part of the contractor.

128. In a cost-plus contract, the contractor is paid for:
a. All the costs incurred plus a fixed fee.
b. Labor costs plus a percentage of the profit.
c. Material costs only plus a fee.
d. Direct costs plus indirect costs.

Answer: a. All the costs incurred plus a fixed fee.
Explanation: In a cost-plus contract, the owner agrees to pay the contractor for all incurred costs, direct and indirect, plus a fixed or percentage fee.

129. A contractor and a property owner disagree on the interpretation of a contract term. This scenario highlights the importance of:
a. Ambiguity resolution clauses.
b. Liquidated damages clauses.
c. Performance bonds.
d. Unilateral contract modifications.

Answer: a. Ambiguity resolution clauses.
Explanation: Ambiguity resolution clauses or clear and unambiguous contract terms are crucial to prevent disagreements on the interpretation of contract terms.

130. The contractor receives a "Notice to Proceed" but there is an unforeseen delay due to weather conditions. The contractor should:
a. Continue the work as per schedule irrespective of the conditions.
b. Notify relevant parties and negotiate a revised schedule.
c. Claim additional costs incurred due to delays.
d. Wait for clear weather and then proceed without any formal notification.

Answer: b. Notify relevant parties and negotiate a revised schedule.
Explanation: When unforeseen conditions cause delays, contractors should communicate with relevant parties and may need to negotiate revised schedules to maintain contractual compliance.

131. When a contractor fails to pay overtime to eligible workers, they are primarily violating:
a. The Occupational Safety and Health Act.
b. The Equal Pay Act.
c. The Fair Labor Standards Act.
d. The National Labor Relations Act.

Answer: c. The Fair Labor Standards Act. Explanation: The Fair Labor Standards Act (FLSA) regulates issues related to employee wages, including overtime pay, ensuring eligible employees are compensated accordingly for hours worked in excess of the standard workweek.

132. In a scenario where an employee reports unsafe working conditions, the contractor is obligated under OSHA to:
a. Dismiss the employee's concerns unless they are shared by others.
b. Address the concerns and rectify any verified unsafe conditions.
c. Transfer the employee to a different project.
d. Report the employee to higher management for disruption.

Answer: b. Address the concerns and rectify any verified unsafe conditions. Explanation: OSHA mandates that employers must maintain a safe working environment and promptly address and rectify any reported unsafe working conditions, safeguarding employees' well-being.

133. In situations where a subcontractor's employee sustains an injury at the worksite, the prime contractor:
a. Bears no responsibility, as the employee is not directly hired.
b. Is solely responsible for all medical costs and compensations.
c. Must share the liability with the subcontractor, dividing the costs equally.
d. Might bear some responsibility, depending on contractual and legal provisions.

Answer: d. Might bear some responsibility, depending on contractual and legal provisions.
Explanation: The responsibility may fall partially or fully on the prime contractor, contingent on the established contractual agreements, legal stipulations, and specific circumstances surrounding the incident.

134. If an employee is dismissed for reporting illegal activities within the organization (whistleblowing), this dismissal is considered:
a. Legitimate, as it protects company reputation.
b. Unlawful and retaliatory under whistleblower protection laws.
c. Conditional, based on the severity of the reported activity.
d. Valid, if the reported activity is unproven.

Answer: b. Unlawful and retaliatory under whistleblower protection laws.
Explanation: Whistleblower protection laws shield employees who expose illegal activities within the organization from retaliatory actions, including unlawful dismissal, ensuring the protection of employees acting in good faith.

135. To avoid disputes related to employee classification, a contractor must:
a. Designate all workers as independent contractors.
b. Classify workers based on their preference.
c. Accurately differentiate between employees and independent contractors based on legal criteria.
d. Rotate workers between classifications periodically.

Answer: c. Accurately differentiate between employees and independent contractors based on legal criteria.
Explanation: Correct classification, based on specific legal criteria, is crucial in ensuring compliance with employment and labor laws, impacting tax obligations, benefits, and legal protections.

136. If a contractor implements a policy violating workers' right to discuss their pay with colleagues, they are infringing upon provisions of:
a. The Equal Pay Act.
b. The National Labor Relations Act.
c. The Fair Labor Standards Act.
d. The Occupational Safety and Health Act.

Answer: b. The National Labor Relations Act.
Explanation: The National Labor Relations Act (NLRA) protects employees' rights to engage in concerted activities, including discussing wages, working conditions, and employment terms, prohibiting employer policies restricting such discussions.

137. When a contractor deducts uniform costs from an employee's paycheck, rendering their pay below minimum wage, this act is non-compliant with:
a. The Davis-Bacon Act.
b. The Service Contract Act.
c. The Fair Labor Standards Act.
d. The Walsh-Healey Public Contracts Act.

Answer: c. The Fair Labor Standards Act.
Explanation: The Fair Labor Standards Act (FLSA) mandates adherence to minimum wage standards, rendering any deductions resulting in pay below the minimum wage as non-compliant with the law.

138. In a project involving federal funds, a contractor must ensure wage compliance with:
a. The McNamara-O'Hara Service Contract Act.
b. The Davis-Bacon Act.
c. The Walsh-Healey Public Contracts Act.
d. The Contract Work Hours and Safety Standards Act.

Answer: b. The Davis-Bacon Act.
Explanation: The Davis-Bacon Act stipulates that contractors involved in federally funded construction projects must pay laborers and mechanics no less than the locally prevailing wages and fringe benefits.

139. A contractor, by displaying a poster detailing employees' rights under the Fair Labor Standards Act, is fulfilling:
a. A recommended best practice.
b. An optional employee incentive.
c. A mandatory compliance requirement.
d. An organizational transparency initiative.

Answer: c. A mandatory compliance requirement.
Explanation: Employers are obligated under the FLSA to display an official poster outlining the provisions of the act, ensuring employee awareness of their rights and entitlements.

140. When a contractor fails to provide a safe and healthful workplace, they are in violation of:
a. The Davis-Bacon Act.
b. The Walsh-Healey Public Contracts Act.
c. The Occupational Safety and Health Act.
d. The Fair Labor Standards Act.

Answer: c. The Occupational Safety and Health Act.
Explanation: The Occupational Safety and Health Act (OSHA) mandates that employers are responsible for providing a safe and healthful workplace, safeguarding employees against hazards.

141. To preserve lien rights, a Preliminary Notice must be served within a specific timeframe after commencing work or delivering materials. The period is typically:
a. Within 5 days
b. Within 20 days
c. Within 30 days
d. Within 45 days

Answer: b. Within 20 days

Explanation: To preserve lien rights, a Preliminary Notice typically needs to be served within 20 days after beginning work or delivering materials, ensuring the involved parties are informed about the claims on the property.

142. In a case where a property owner pays the general contractor but the subcontractor is not paid, the subcontractor can:

a. File a lien against the owner's property.

b. Seek payment only from the general contractor.

c. Do nothing as the contract is with the general contractor.

d. File a legal complaint against the property owner.

Answer: a. File a lien against the owner's property.

Explanation: Even if the property owner has paid the general contractor, a subcontractor can still file a lien against the owner's property if they haven't been paid, protecting their right to compensation.

143. When a contractor receives a stop payment notice from a subcontractor, the contractor must:

a. Continue working and ignore the notice.

b. Freeze any payment from the owner relating to the subcontractor's work.

c. Pay the subcontractor immediately.

d. File a lawsuit against the subcontractor.

Answer: b. Freeze any payment from the owner relating to the subcontractor's work.

Explanation: A stop payment notice requires the contractor to freeze any related payments from the owner until the dispute with the subcontractor is resolved, safeguarding the subcontractor's claim.

144. In a scenario where a mechanic's lien is placed on a property, the property can be:

a. Sold as is.

b. Auctioned off to satisfy the lien.

c. Leased without any restrictions.

d. Temporarily seized.

Answer: b. Auctioned off to satisfy the lien.

Explanation: If a mechanic's lien is placed on a property and remains unsatisfied, the property can be auctioned off to satisfy the amount owed under the lien, impacting the owner's property rights.

145. To successfully enforce a mechanic's lien, a contractor must typically file a lawsuit within:

a. 30 days after recording the lien.

b. 60 days after recording the lien.

c. 90 days after recording the lien.

d. 120 days after recording the lien.

Answer: c. 90 days after recording the lien.
Explanation: A contractor typically has 90 days after recording a mechanic's lien to file a lawsuit to enforce the lien, failure to do so may result in the lien becoming null and void.

146. When a lien release is signed and the contractor has not yet been paid:
a. The lien remains effective until payment is received.
b. The contractor forfeits the right to the lien.
c. The lien can be refiled after payment is received.
d. The lien automatically converts to a lawsuit.

Answer: b. The contractor forfeits the right to the lien.
Explanation: Signing a lien release before receiving payment usually means forfeiting the right to the lien, risking non-payment without any lien protection.

147. When multiple liens are placed on a property, the priority of liens is generally determined by:
a. The amount of the lien.
b. The date the lien is recorded.
c. The status of the lienor.
d. The type of work performed.

Answer: b. The date the lien is recorded.
Explanation: Generally, the priority of liens is determined by the date the lien is recorded, meaning liens recorded earlier have higher priority over those recorded later.

148. In a project where laborers have not been paid and file a lien, this lien is usually:
a. Subordinated to material suppliers' liens.
b. Prioritized over other liens.
c. Treated equally with all other liens.
d. Invalidated if the contractor is paid.

Answer: b. Prioritized over other liens.
Explanation: Typically, laborers' liens, due to the nature of labor being a critical component of construction projects, are given priority over other types of liens to ensure protection of laborers' rights.

149. When a Notice of Completion is recorded on a project, the timeframe for filing a mechanic's lien is:
a. Extended indefinitely.
b. Shortened significantly.
c. Unaffected.
d. Extended by 30 days.

Answer: b. Shortened significantly.
Explanation: Recording a Notice of Completion generally shortens the timeframe for filing a mechanic's lien, necessitating prompt action by those wishing to file a lien to secure their claims.

150. A subcontractor who has filed a mechanic's lien can typically enforce their claim by:
a. Mediating with the property owner.
b. Arbitrating with the general contractor.
c. Filing a lawsuit to foreclose on the lien.
d. Submitting a claim to the contractor's bond.

Answer: c. Filing a lawsuit to foreclose on the lien.
Explanation: To enforce a mechanic's lien, a subcontractor typically needs to file a lawsuit to foreclose on the lien, moving the legal process forward to claim the owed amount.

151. In a construction project scenario, if an employer fails to provide fall protection, OSHA may impose penalties. What is the max penalty per violation that OSHA can impose?
a. $5,000
b. $13,653
c. $25,000
d. $50,000

Answer: b. $13,653
Explanation: OSHA can impose a penalty of up to $13,653 per violation for serious, other-than-serious, and posting requirements violations.

152. While working on a construction site, workers are required to wear hard hats when there is a potential for head injury from impacts. Hard hats must be:
a. ANSI Z89.1 compliant
b. ISO 45001 certified
c. NFPA 70E rated
d. ASTM F2413 standard

Answer: a. ANSI Z89.1 compliant
Explanation: Hard hats must meet the standards outlined in ANSI Z89.1, indicating compliance with the requirements for impact and penetration resistance.

153. When an incident occurs at a construction site resulting in hospitalization, the employer must report the incident to OSHA within:
a. 8 hours
b. 24 hours
c. 48 hours
d. 72 hours

Answer: b. 24 hours
Explanation: Employers are required to report any in-patient hospitalization to OSHA within 24 hours of the incident.

154. A job site has multiple contractors and subcontractors working simultaneously. Who is primarily responsible for ensuring the health and safety of all workers?
a. The general contractor
b. Each individual contractor
c. The project manager
d. The site safety officer

Answer: a. The general contractor
Explanation: The general contractor is primarily responsible for maintaining a safe and healthy work environment for all workers on the site, regardless of their employer.

155. OSHA's fall protection standards stipulate that fall protection is required when working above:
a. 4 feet
b. 6 feet
c. 10 feet
d. 15 feet

Answer: b. 6 feet
Explanation: OSHA mandates that workers must be provided with fall protection when working at heights of 6 feet or more above a lower level.

156. A scaffold used in construction must be capable of supporting, without failure, its own weight and at least how many times the intended load?
a. 2 times
b. 3 times
c. 4 times
d. 5 times

Answer: c. 4 times
Explanation: Scaffolds must be designed to support their own weight and at least four times the intended load to ensure stability and safety.

157. A worker is using a portable ladder to access an elevated work area. What is the minimum distance the ladder must extend above the landing?
a. 1 foot
b. 2 feet
c. 3 feet
d. 4 feet

Answer: c. 3 feet
Explanation: Ladders must extend at least 3 feet above the landing to provide a sufficient handhold for safely mounting and dismounting.

158. During excavation, a trench that is 5 feet deep or more requires a protective system unless it is made entirely in stable rock. What is the maximum allowable slope for a trench in Type A soil?
a. 45 degrees
b. 53 degrees
c. 63 degrees
d. 75 degrees

Answer: b. 53 degrees
Explanation: For Type A soil, the maximum allowable slope for a trench is 53 degrees from the horizontal, ensuring stability and preventing cave-ins.

159. When employees are exposed to hazardous substances, they must receive training in accordance with:
a. HazCom Standard
b. Lockout/Tagout Standard
c. Machine Guarding Standard
d. Fall Protection Standard

Answer: a. HazCom Standard
Explanation: The Hazard Communication Standard (HazCom) requires employee training on hazardous substances to ensure that they are aware of the potential risks and know how to protect themselves.

160. Regarding noise exposure, employees must be included in a hearing conservation program if they are exposed to an average noise level of:
a. 80 dBA or above
b. 85 dBA or above
c. 90 dBA or above
d. 95 dBA or above

Answer: b. 85 dBA or above
Explanation: Employees exposed to an average noise level of 85 decibels A-weighted (dBA) or above must be included in a hearing conservation program to prevent hearing loss due to high noise levels.

161. A construction company is planning to develop a multi-story commercial building in a zone marked as "residential" in the city's zoning map. What step must the company take first?
a. Start construction immediately.
b. Apply for a zoning variance.
c. Appeal to the building department.
d. Purchase building permits.

Answer: b. Apply for a zoning variance.
Explanation: Construction companies must apply for a zoning variance to get approval for construction that doesn't conform to the zoning designation of the area.

162. During a commercial building inspection, the inspector notices a violation of the local building code. What is the usual next step?
a. Issuance of a demolition order.
b. Issuance of a stop-work order.
c. Immediate rectification by the inspector.
d. Suspension of the contractor's license.

Answer: b. Issuance of a stop-work order.
Explanation: Typically, the issuance of a stop-work order is the next step, halting construction until the violation is addressed and resolved.

163. In a building project, the fire protection systems are required to comply with which code?
a. International Building Code (IBC)
b. International Fire Code (IFC)
c. International Residential Code (IRC)
d. International Energy Conservation Code (IECC)

Answer: b. International Fire Code (IFC)
Explanation: Fire protection systems are regulated by the International Fire Code (IFC) which sets the standards for fire safety in buildings.

164. If a contractor wishes to build closer to the property line than allowed by zoning ordinances, they must seek:
a. Rezoning
b. Special-use permit
c. Variance
d. Conditional use permit

Answer: c. Variance
Explanation: A variance is needed when a contractor wants to deviate from the set zoning requirements, such as building closer to the property line.

165. When a city updates its building codes, existing structures are:
a. Grandfathered under the code in effect when they were built.
b. Required to be updated immediately.
c. Given a 5-year grace period to comply.
d. Assessed and selectively enforced.

Answer: a. Grandfathered under the code in effect when they were built.
Explanation: Typically, existing structures are grandfathered under the code that was in effect at the time they were built and are not obliged to comply with newer codes unless substantial renovations are made.

166. A contractor is working on a structure in a flood-prone area. The construction should comply with:
a. Zoning ordinances only.
b. Local building codes only.
c. Both zoning ordinances and local building codes.
d. Neither zoning ordinances nor local building codes.

Answer: c. Both zoning ordinances and local building codes.
Explanation: In flood-prone areas, constructions must comply with local zoning ordinances related to land use and local building codes to ensure structural safety and resilience.

167. A building built in compliance with the building codes of 2005 needs renovations. The city has adopted the 2021 International Building Code (IBC). The renovations must comply with:
a. The 2005 Building Codes.
b. The 2021 International Building Code (IBC).
c. A mix of both 2005 and 2021 codes.
d. The choice is at the discretion of the contractor.

Answer: b. The 2021 International Building Code (IBC).
Explanation: Any new renovations or constructions must comply with the current adopted building codes, which in this scenario is the 2021 IBC.

168. A zoning ordinance primarily regulates:
a. Structural integrity of the buildings.
b. Use, density, and form of the land.
c. Fire safety measures in a building.
d. Plumbing and electrical systems in a building.

Answer: b. Use, density, and form of the land.
Explanation: Zoning ordinances are primarily concerned with regulating the use, density, and form of land within a jurisdiction to coordinate land use and development.

169. A contractor receives a project in an area he hasn't worked before. Which of the following should he refer to first to understand the building requirements and restrictions?
a. The local zoning ordinances.
b. The International Building Code (IBC).
c. The project's blueprints.
d. The National Electrical Code (NEC).

Answer: a. The local zoning ordinances.
Explanation: Before starting the project, the contractor should refer to local zoning ordinances to understand the land use, building requirements, and restrictions in the area.

170. In a case where local building codes conflict with international building codes, which takes precedence?
a. International Building Code (IBC)
b. Local Building Code
c. A consensus between the two
d. The contractor's discretion

Answer: b. Local Building Code
Explanation: Local building codes take precedence over international codes as they are tailored to address the specific needs, conditions, and priorities of a locality.

171. A contractor is working on a project located near a wetland area. What environmental regulation would typically apply to avoid impacts on this sensitive habitat?
a. Clean Air Act
b. Clean Water Act
c. Endangered Species Act
d. Resource Conservation and Recovery Act

Answer: b. Clean Water Act
Explanation: The Clean Water Act governs water pollution and would typically apply to protect sensitive habitats such as wetlands from project impacts.

172 A construction company is planning a project in an area known to be habitat for an endangered species. What is the primary environmental regulation that needs to be considered?
a. Clean Air Act
b. Clean Water Act
c. Endangered Species Act
d. National Environmental Policy Act

Answer: c. Endangered Species Act
Explanation: The Endangered Species Act protects the habitats of species that are determined to be endangered or threatened.

173. When managing construction waste, which act provides guidelines regarding the disposal of hazardous waste?
a. Clean Air Act
b. Clean Water Act
c. Resource Conservation and Recovery Act
d. Comprehensive Environmental Response, Compensation, and Liability Act

Answer: c. Resource Conservation and Recovery Act
Explanation: The Resource Conservation and Recovery Act (RCRA) gives EPA the authority to control hazardous waste from the "cradle-to-grave."

174. To ensure compliance with stormwater discharge regulations on a construction site, contractors primarily refer to which law?
a. Clean Air Act
b. Clean Water Act
c. Resource Conservation and Recovery Act
d. Endangered Species Act

Answer: b. Clean Water Act
Explanation: The Clean Water Act provides the framework for regulating discharges of pollutants, including stormwater discharges, into the waters of the United States.

175. Which environmental law would require a contractor to evaluate the environmental impacts of a large construction project before it begins?
a. Clean Water Act
b. National Environmental Policy Act (NEPA)
c. Endangered Species Act
d. Clean Air Act

Answer: b. National Environmental Policy Act (NEPA)
Explanation: NEPA requires federal agencies to assess the environmental impacts of their proposed actions prior to making decisions.

176. A contractor working on a residential project discovers an underground storage tank leaking an unknown substance. What is the contractor's immediate legal obligation?
a. Remove the tank.
b. Notify the local environmental agency.
c. Continue work and monitor the situation.
d. Seal the tank and assess the damage.

Answer: b. Notify the local environmental agency.
Explanation: The immediate legal obligation is to report the discovery to the appropriate environmental agency to assess the situation and determine subsequent steps.

177. A contractor wishes to use a chemical substance for pest control during construction. Which agency should the contractor refer to, for restrictions and approved substances?
a. Environmental Protection Agency (EPA)
b. Occupational Safety and Health Administration (OSHA)
c. National Institute for Occupational Safety and Health (NIOSH)
d. United States Geological Survey (USGS)

Answer: a. Environmental Protection Agency (EPA)
Explanation: The EPA regulates pesticides under the Federal Insecticide, Fungicide, and Rodenticide Act (FIFRA) to ensure they do not pose adverse effects on humans or the environment.

178. A contractor is working on a renovation project in a building constructed in the 1960s. The contractor is likely to encounter which hazardous material that is regulated by environmental laws?
a. Lead
b. Mercury
c. Asbestos
d. Chlorofluorocarbons

Answer: c. Asbestos
Explanation: Buildings constructed before the 1970s are likely to contain asbestos, a hazardous material, and its removal and disposal are strictly regulated.

179. When constructing a new building, which regulation dictates the standards for energy efficiency and environmental design?
a. Energy Policy Act
b. Clean Air Act
c. Resource Conservation and Recovery Act
d. National Environmental Policy Act (NEPA)

Answer: a. Energy Policy Act
Explanation: The Energy Policy Act sets forth energy efficiency standards and guidelines for different sectors including building construction, to conserve energy and reduce environmental impact.

180. In a construction project that will alter the landscape, which regulatory document typically outlines the necessary measures to mitigate soil erosion?
a. Environmental Impact Statement
b. Stormwater Pollution Prevention Plan (SWPPP)
c. Spill Prevention, Control, and Countermeasure (SPCC) Plan
d. Resource Management Plan

Answer: b. Stormwater Pollution Prevention Plan (SWPPP)
Explanation: An SWPPP is a fundamental requirement that outlines the measures to prevent stormwater discharges and control soil erosion and sediment in construction projects affecting the landscape.

181. A masonry contractor is building a brick wall and needs to select a suitable mortar. Which type of mortar is most appropriate for a wall that requires high compressive strength?
a. Type N Mortar
b. Type S Mortar
c. Type O Mortar
d. Type M Mortar

Answer: d. Type M Mortar
Explanation: Type M mortar has the highest compressive strength, making it suitable for constructions requiring high strength like foundations and retaining walls.

182. When installing a metal roof on a commercial building, what is the most crucial consideration to avoid future corrosion?
a. Slope of the roof
b. Proximity to the ocean
c. Type of metal used
d. Thickness of the metal

Answer: c. Type of metal used
Explanation: The type of metal used is critical to prevent corrosion, especially in environments with corrosive elements like salt air.

183. A contractor is tasked with installing HVAC systems in a high-rise building. For energy efficiency and minimal heat loss, which type of ductwork insulation is most appropriate?
a. Fiberglass insulation
b. Reflective insulation
c. Spray foam insulation
d. Mineral wool insulation

Answer: c. Spray foam insulation
Explanation: Spray foam insulation is known for its high R-value and ability to prevent air leaks, making it suitable for energy-efficient HVAC installations.

184. An electrician is wiring a new residential construction project. Which wire color is universally used to represent the ground wire?
a. Red
b. Black
c. White
d. Green or bare

Answer: d. Green or bare
Explanation: In electrical wiring, green or bare wires are universally used to represent the ground wire.

185. For a plumbing contractor, what is the primary consideration when selecting pipe material for a residential potable water supply system?
a. Cost of material
b. Resistance to corrosion
c. Ease of installation
d. Aesthetic appeal

Answer: b. Resistance to corrosion
Explanation: Resistance to corrosion is crucial to ensure the longevity and safety of the potable water supply system, preventing contamination and leaks.

186. A contractor is installing windows in a coastal area prone to hurricanes. Which type of glass is mandated by building codes in such regions to withstand high winds and flying debris?
a. Annealed glass
b. Tempered glass
c. Laminated glass
d. Insulated glass

Answer: c. Laminated glass
Explanation: Laminated glass is designed to hold together when shattered, providing an essential level of protection against windborne debris in hurricane-prone areas.

187. A flooring contractor is installing tiles in a high-traffic area. Which type of tile would be best suited for areas expecting heavy footfall and intense use?
a. Ceramic tiles
b. Porcelain tiles
c. Vinyl tiles
d. Cork tiles

Answer: b. Porcelain tiles
Explanation: Porcelain tiles are denser, more durable, and less porous than ceramic tiles, making them suitable for high-traffic areas.

188. For a commercial project requiring soundproofing between offices, which type of drywall would a contractor most likely use?
a. Regular drywall
b. Moisture-resistant drywall
c. Fire-resistant drywall
d. Soundproof drywall

Answer: d. Soundproof drywall
Explanation: Soundproof drywall is specifically designed to reduce noise transmission between rooms and would be the most suitable choice for soundproofing offices.

189. A contractor needs to select a suitable paint for a bathroom project. What type of paint finish is most appropriate for areas with high humidity and moisture?
a. Flat finish
b. Eggshell finish
c. Satin finish
d. Semi-gloss finish

Answer: d. Semi-gloss finish
Explanation: Semi-gloss finish is resistant to humidity and moisture, making it suitable for bathrooms, and it's easier to clean compared to other finishes.

190. A construction project requires a concrete mix with high workability and placement ease. Which admixture should the contractor use to achieve the desired properties in the concrete mix?
a. Accelerating admixture
b. Retarding admixture
c. Water-reducing admixture
d. Air-entraining admixture

Answer: c. Water-reducing admixture
Explanation: Water-reducing admixtures increase the workability of the concrete mix without changing the water content, facilitating ease of placement.

191. A contractor is building a retaining wall and needs to determine the appropriate type of reinforcement. Which factor is most crucial in deciding the reinforcement method?
a. Aesthetic preferences
b. Soil type behind the wall
c. Proximity to vegetation
d. Wall height

Answer: b. Soil type behind the wall
Explanation: The type of soil behind the wall is crucial in deciding the reinforcement as it directly impacts the pressure exerted on the wall and hence, the strength required to retain it.

192. A construction project involves erecting a steel frame for a commercial building. Which welding technique is most suitable for joining large steel beams?
a. Spot welding
b. Arc welding
c. Gas welding
d. Laser welding

Answer: b. Arc welding
Explanation: Arc welding is commonly used for joining large steel structures as it provides strong and durable welds suitable for supporting substantial loads.

193. When constructing a residential building in a flood-prone area, which foundation type is most suitable to minimize flood damage?
a. Slab-on-grade foundation
b. Pier and beam foundation
c. Full basement foundation
d. Crawl space foundation

Answer: b. Pier and beam foundation
Explanation: Pier and beam foundations elevate the house, reducing the risk of floodwater entering living spaces and causing damage.

194. During the construction of a high-rise building, a contractor opts for a post-tensioned concrete slab system. What is the primary benefit of this choice?
a. Reduced concrete cracking
b. Enhanced aesthetic appeal
c. Increased fire resistance
d. Improved thermal insulation

Answer: a. Reduced concrete cracking
Explanation: Post-tensioned slabs allow for longer spans and reduce the occurrence of cracks in the concrete, providing durability and structural integrity.

195. A contractor needs to select the appropriate material for the exterior siding of a house located in an area with extreme temperature fluctuations. Which material is best suited for this environment?
a. Vinyl siding
b. Wood siding
c. Brick veneer
d. Metal siding

Answer: c. Brick veneer
Explanation: Brick veneer is durable and can withstand extreme temperature fluctuations, making it suitable for areas with varying climates.

196. For a green building project, a contractor is evaluating insulation options. Which insulation material offers the highest R-value per inch?
a. Fiberglass
b. Cellulose
c. Closed-cell spray foam
d. Mineral wool

Answer: c. Closed-cell spray foam
Explanation: Closed-cell spray foam has the highest R-value per inch among the listed options, making it the most effective insulator.

197. During the construction of a commercial building, a contractor encounters expansive clay soil on the site. Which type of foundation is most suitable for such soil conditions?
a. Mat foundation
b. Strip foundation
c. Raft foundation
d. Pile foundation

Answer: d. Pile foundation
Explanation: Pile foundation is suitable for expansive clay soils as it transfers loads to stable strata below the soil, preventing structural damage due to soil movement.

198. A contractor is planning to install a HVAC system in a residential building. Which type of ductwork is best suited for areas with space constraints?
a. Flexible ducts
b. Spiral ducts
c. Rectangular ducts
d. Oval ducts

Answer: a. Flexible ducts
Explanation: Flexible ducts can be routed around obstacles, making them suitable for areas where space is constrained.

199. When constructing a residential building, a contractor must select an appropriate roofing material. Which material is best suited for a house located in a region prone to wildfires?
a. Wood shakes
b. Asphalt shingles
c. Clay tiles
d. Rubber slate

Answer: c. Clay tiles
Explanation: Clay tiles are non-combustible and offer excellent fire resistance, making them suitable for areas prone to wildfires.

200. A contractor is installing flooring in a commercial kitchen. Which flooring material is best suited to handle frequent spillages and heavy foot traffic?
a. Laminate
b. Hardwood
c. Ceramic tile
d. Carpet

Answer: c. Ceramic tile
Explanation: Ceramic tile is durable, water-resistant, and easy to clean, making it ideal for commercial kitchens where spillages are frequent, and traffic is heavy.

201. A contractor is analyzing a complex set of blueprints for a multifamily residential project. What does a dashed line typically represent on a floor plan?
a. Permanent structures
b. Above-ceiling elements
c. Proposed construction
d. Load-bearing walls

Answer: b. Above-ceiling elements
Explanation: In blueprint reading, dashed lines typically represent elements that are above the ceiling, such as overhead beams or utilities.

202. While reviewing the construction documents for a commercial building project, a contractor notices a discrepancy between the plumbing and electrical plans. Which course of action is most appropriate?
a. Ignore the discrepancy
b. Resolve the discrepancy by consulting with the subcontractors
c. Modify the blueprints without consultation
d. Consult the project architect or engineer to resolve the discrepancy

Answer: d. Consult the project architect or engineer to resolve the discrepancy
Explanation: Any discrepancy in construction documents should be addressed by consulting with the project architect or engineer to ensure that modifications are accurate and conform to design specifications.

203. A contractor is studying the elevations on a set of residential blueprints. Which information can typically be derived from the elevations?
a. Electrical layouts
b. Exterior finishes
c. Plumbing fixtures
d. HVAC ductwork

Answer: b. Exterior finishes
Explanation: Elevations primarily provide information on the exterior finishes, height dimensions, and the appearance of the building from each side.

204. During the plan analysis for a new office building, a contractor needs to determine the building's orientation. Which document is crucial for finding this information?
a. Floor plan
b. Site plan
c. Electrical plan
d. Plumbing plan

Answer: b. Site plan
Explanation: The site plan provides information on the building's orientation, property lines, and the location of the building on the site.

205. A contractor is interpreting a structural plan for a high-rise project. What does a symbol with parallel lines represent in structural drawings?
a. Reinforcing bars
b. Steel beams
c. Concrete walls
d. Masonry units

Answer: a. Reinforcing bars
Explanation: In structural drawings, parallel lines typically represent reinforcing bars (rebar) used to strengthen concrete structures.

206. While analyzing a set of blueprints for a residential project, a contractor needs to identify the type and location of light fixtures. Which type of drawing will provide this information?
a. Floor plan
b. Electrical plan
c. Reflected ceiling plan
d. Elevation drawing

Answer: c. Reflected ceiling plan
Explanation: A reflected ceiling plan provides details about the ceiling, including the location and type of light fixtures, HVAC vents, and sprinkler heads.

207. A contractor is reviewing the blueprints of a commercial building. To understand the slope and height of the roof, which drawing should be referred to?
a. Section drawing
b. Site plan
c. Elevation drawing
d. Floor plan

Answer: a. Section drawing
Explanation: Section drawings cut through the building and are used to show details of the construction, including the slope and height of the roof.

208. While examining a set of construction plans for a residential building, a contractor must ascertain the type and location of windows. Which drawing typically provides this detailed information?
a. Floor plan
b. Section drawing
c. Elevation drawing
d. Detail drawing

Answer: d. Detail drawing
Explanation: Detail drawings are used to provide comprehensive information about specific components of a project, such as windows, including their type, location, and installation details.

209. A contractor is studying a residential project's blueprints. What is the significance of a cloud drawn around a portion of the plan?
a. It highlights a structural concern
b. It indicates an area with HVAC requirements
c. It signifies a revision or modification
d. It represents an area with high moisture

Answer: c. It signifies a revision or modification
Explanation: In blueprint reading, a cloud around a portion of the plan typically denotes a recent revision or modification made to that area.

210. A contractor is interpreting blueprints for a commercial project. The plans specify using 'S' type joints in a particular section. What does 'S' represent in this context?
a. Steel joint
b. Soldered joint
c. Slip joint
d. Seismic joint

Answer: d. Seismic joint
Explanation: In construction blueprints, an 'S' type joint typically refers to a Seismic joint designed to accommodate movement between building sections in the event of an earthquake.

211. A contractor is considering various materials for the exterior of a coastal residence. Which material would be best suited to resist the corrosive effects of a salt-laden environment?
a. Aluminum siding
b. Mild steel panels
c. Copper cladding
d. Galvanized steel siding

Answer: c. Copper cladding
Explanation: Copper cladding is highly resistant to corrosion from salt and other environmental conditions, making it ideal for coastal applications. It forms a protective patina that safeguards it from corrosion.

212. During the construction of a hospital, the contractor needs to select a flooring material that is durable, easy to clean, and inhibits microbial growth. Which of the following materials is most suitable?
a. Hardwood
b. Carpet
c. Epoxy resin
d. Porcelain tile

Answer: c. Epoxy resin. Explanation: Epoxy resin flooring is known for its durability, ease of maintenance, and is often used in medical facilities due to its resistance to bacteria and other microbes.

213. In a residential construction project, a contractor is determining the best insulation material to maintain energy efficiency and sound insulation. Which insulation material would be most effective for this purpose?
a. Fiberglass insulation
b. Cellulose insulation
c. Spray foam insulation
d. Mineral wool insulation

Answer: c. Spray foam insulation
Explanation: Spray foam insulation has a high R-value, providing superior thermal resistance, and also acts as an effective barrier to air leaks and outside noise.

214. A contractor is working on a project that requires a material with high tensile strength and durability. Which of the following materials is most suitable for such requirements?
a. Aluminum alloy
b. Mild steel
c. High-density polyethylene (HDPE)
d. Reinforced concrete

Answer: d. Reinforced concrete
Explanation: Reinforced concrete has high tensile strength due to the embedded steel bars and is highly durable, making it suitable for structures requiring strength and durability.

215. When selecting materials for a residential roofing project in a region with heavy snowfall, which type of roofing material would best prevent snow accumulation?
a. Flat concrete tiles
b. Metal roofing
c. Asphalt shingles
d. Clay tiles

Answer: b. Metal roofing
Explanation: Metal roofing is preferable in snowy regions as its smooth, slippery surface allows snow to slide off, preventing accumulation.

216. A contractor is selecting a material for a high-traffic flooring area in a commercial building. Which material is best suited for high durability and ease of maintenance?
a. Carpet tiles
b. Laminate flooring
c. Polished concrete
d. Vinyl flooring

Answer: c. Polished concrete
Explanation: Polished concrete is highly durable, resistant to high traffic, and easy to maintain, making it ideal for commercial applications.

217. A contractor needs to select a wall material for a building located in a flood-prone area. Which material would offer the most resilience against water damage and mold growth?
a. Gypsum board
b. Plywood
c. Cement board
d. Fiberboard

Answer: c. Cement board
Explanation: Cement board is water-resistant and will not support mold growth, making it an ideal material for buildings in flood-prone areas.

218. For a project located in a high seismic activity zone, a contractor needs to select a material that can absorb and dissipate energy. Which of the following materials would be the most suitable choice?
a. Brittle concrete
b. Mild steel
c. Glass
d. Base isolators

Answer: d. Base isolators
Explanation: Base isolators are specifically designed to absorb and dissipate seismic energy, enhancing the seismic resilience of structures located in earthquake-prone areas.

219. While selecting windows for a building project aiming to achieve high energy efficiency, a contractor has to choose the type of glazing. Which glazing would best minimize heat transfer?
a. Single glazing
b. Double glazing with low-E coating
c. Double glazing without coating
d. Tinted glazing

Answer: b. Double glazing with low-E coating
Explanation: Double glazing with low-E (low emissivity) coating minimizes heat transfer, reducing heating and cooling costs and improving overall energy efficiency.

220. In a residential project, a contractor is evaluating materials for outdoor decking in a humid climate. Which material is most resistant to rot and insect damage?
a. Pressure-treated lumber
b. Natural cedar
c. Untreated pine
d. Composite decking

Answer: d. Composite decking
Explanation: Composite decking is highly resistant to rot, insects, and other environmental damage, making it a suitable material for outdoor decking in humid climates.

221. A contractor preparing a bid for a commercial construction project needs to consider various direct costs. Which of the following would NOT typically be classified as a direct cost?
a. Labor costs
b. Material costs
c. Equipment rental
d. Office overhead

Answer: d. Office overhead
Explanation: Office overhead is typically classified as an indirect cost as it does not directly contribute to the construction work. Direct costs include labor, materials, and equipment related directly to the project.

222. In a high-end residential construction project, a contractor is utilizing value engineering to optimize the cost. Which approach is MOST likely to produce cost savings without compromising quality?
a. Selecting cheaper, lower-quality materials
b. Reducing the project scope
c. Exploring alternative construction methods
d. Reducing labor costs by hiring less skilled workers

Answer: c. Exploring alternative construction methods
Explanation: Value engineering aims at improving the value of goods or products and services by examining function. Exploring alternative construction methods can lead to cost savings without compromising quality.

223. When estimating the costs for a mixed-use development project, which method would a contractor likely use during the early conceptual design phase?
a. Detailed quantity takeoff method
b. Square-footage method
c. Unit-price method
d. Assembly method

Answer: b. Square-footage method. Explanation: The square-footage method is often used in the early stages of design when detailed information is not available. It provides a rough estimate based on the project's total square footage and average cost per square foot.

224. A contractor receives a request for a proposal that requires a fast-track construction method. What primary risk should the contractor consider when preparing the bid?
a. Reduced construction time
b. Increased material costs
c. Design changes during construction
d. Lower labor costs

Answer: c. Design changes during construction. Explanation: In fast-track construction, design and construction phases can overlap, potentially leading to frequent design changes during construction, affecting cost and schedule.

225. In developing a bid for a residential project, a contractor chooses to add a 10% markup to the total estimated cost to determine the bid price. What is the PRIMARY purpose of this markup?
a. To cover unexpected cost overruns
b. To cover overhead and profit
c. To cover labor costs
d. To cover material costs

Answer: b. To cover overhead and profit
Explanation: The markup is added to cover the contractor's overhead expenses and to ensure a profit margin is realized from the project.

226. For a commercial building project, a contractor is determining the costs associated with concrete works. Which of the following would MOST likely lead to a cost overrun in the concrete scope of work?
a. Lower fuel prices
b. Efficient formwork design
c. Inaccurate quantity takeoff
d. Availability of bulk discounts

Answer: c. Inaccurate quantity takeoff
Explanation: An inaccurate quantity takeoff can lead to understated quantities and therefore underestimated costs, which can result in cost overruns in the concrete scope of work.

227. During the pre-bid phase, a contractor is performing a site analysis for a new suburban development project. Which factor is LEAST likely to impact the cost estimation?
a. Soil condition
b. Site accessibility
c. Nearby competition
d. Availability of utilities

Answer: c. Nearby competition
Explanation: The presence of nearby competition does not directly impact the cost estimation for the project. Factors such as soil condition, site accessibility, and availability of utilities directly impact construction costs.

228. In a competitive bidding situation for a hotel construction project, a contractor decides to adopt a penetration pricing strategy. How is this likely to affect the bid amount?
a. The bid amount will be significantly higher than the estimated costs.
b. The bid amount will be slightly above the estimated costs.
c. The bid amount will be equal to the estimated costs.
d. The bid amount will be lower than the estimated costs.

Answer: d. The bid amount will be lower than the estimated costs.
Explanation: Penetration pricing strategy involves setting a lower price to enter a competitive market and is likely to result in a bid amount that is lower than the estimated costs.

229. While preparing a bid for an office building, a contractor needs to calculate the indirect costs. Which of the following is typically included in the indirect costs?
a. Costs of construction materials
b. Costs of specialized subcontractors
c. Costs of construction equipment
d. Costs of project supervision

Answer: d. Costs of project supervision
Explanation: Indirect costs are those costs not directly attributed to the building process but are nonetheless necessary, such as project supervision, temporary facilities, and utilities.

230. A contractor is preparing a cost estimate for a renovation project and needs to assess labor costs. Which of the following factors is LEAST likely to impact labor cost estimation?
a. The local availability of skilled labor
b. The productivity of the labor force
c. The cost of construction materials
d. Prevailing wage rates

Answer: c. The cost of construction materials
Explanation: The cost of construction materials does not directly impact the estimation of labor costs. Labor costs are more directly influenced by availability, productivity, and prevailing wage rates.

231. When utilizing a pneumatic nail gun for framing purposes, what primary precaution should be taken to avoid tool-related accidents?
a. Inspecting the air hose for kinks and damage
b. Using the tool in a well-ventilated area
c. Using nails of the correct length and gauge
d. Avoiding the use of protective equipment

Answer: a. Inspecting the air hose for kinks and damage
Explanation: Proper inspection of the air hose is crucial to avoid any accidents, as any damage can lead to a loss of pressure or sudden hose whipping, causing serious harm.

232. For a complex commercial construction project, a contractor needs to ensure optimal utilization of a crane. What is the MOST important factor affecting crane productivity?
a. Type of crane used
b. Height of the building
c. Experience of the crane operator
d. Weather conditions

Answer: c. Experience of the crane operator
Explanation: An experienced operator can significantly optimize crane productivity, ensuring smooth operations, efficient material handling, and adherence to safety protocols, thereby reducing downtime.

233. When selecting a circular saw for cutting a variety of materials on-site, what aspect is critical to ensuring versatile and efficient operation?
a. The color of the saw
b. The brand of the saw
c. The availability of compatible blades
d. The weight of the saw

Answer: c. The availability of compatible blades
Explanation: Having a variety of compatible blades allows the tool to be used for cutting different materials efficiently, making it versatile and adaptable to various tasks on-site.

234. A contractor needs to assess the optimal load capacity for an excavator in a residential construction project. Which of the following factors is LEAST relevant in making this determination?
a. Type of soil
b. Excavator's arm reach
c. Proximity to residential structures
d. Manufacturer's specifications

Answer: c. Proximity to residential structures
Explanation: Proximity to residential structures doesn't directly impact the load capacity of an excavator. Manufacturer's specifications, type of soil, and excavator's arm reach are more relevant in determining the optimal load capacity.

235. For a road construction project, which piece of equipment is crucial for compacting granular soils and crushed aggregate?
a. Bulldozer
b. Vibratory roller
c. Backhoe
d. Grader

Answer: b. Vibratory roller
Explanation: Vibratory rollers are specifically designed to compact different types of soil and aggregates, ensuring proper compaction and stability of the road base.

236. When dealing with underground utility installations, what piece of equipment is MOST suited for trenching operations?
a. Skid steer loader
b. Trencher
c. Excavator
d. Pneumatic drill

Answer: b. Trencher
Explanation: Trenchers are specifically designed for creating trenches, especially for the installation of underground utilities, making them the most suitable equipment for such operations.

237. A contractor plans to use a hydraulic breaker for demolishing a concrete structure. Which factor is critical in selecting the appropriate hydraulic breaker?
a. The color of the breaker
b. The noise level of the breaker
c. The breaker's impact energy and frequency
d. The availability of spare parts

Answer: c. The breaker's impact energy and frequency
Explanation: The impact energy and frequency of a hydraulic breaker determine its breaking capability and efficiency, thus are critical in selecting an appropriate breaker for specific demolition tasks.

238. In a high-rise construction project, what piece of equipment is indispensable for transporting personnel and materials to various floor levels?
a. Tower crane
b. Scissor lift
c. Construction elevator
d. Forklift

Answer: c. Construction elevator
Explanation: A construction elevator is essential for efficiently transporting personnel, tools, and materials to different levels of a high-rise building during construction.

239. A contractor must decide on a suitable piece of equipment for moving heavy construction materials across a site. Which of the following would be the MOST appropriate choice for this task?
a. Wheel loader
b. Excavator
c. Boom lift
d. Scaffolding

Answer: a. Wheel loader
Explanation: Wheel loaders are designed to move heavy loads across construction sites efficiently, making them the most suitable choice for transporting construction materials.

240. In a scenario where a contractor needs to perform accurate horizontal and vertical measurements for a large-scale project, which tool is MOST suitable to ensure precision and efficiency?
a. Manual level
b. Plumb bob
c. Laser level
d. Tape measure

Answer: c. Laser level
Explanation: Laser levels provide high precision in both horizontal and vertical measurements over long distances, making them indispensable for large-scale projects requiring exact measurements.

241. In implementing a Quality Management System (QMS) for a construction project, what is the MAIN objective of conducting regular quality audits?
a. To find faults in employees' work habits.
b. To assess compliance with planned arrangements and the effectiveness of the QMS.
c. To fulfill contractual obligations.
d. To avoid liability in the event of project failure.

Answer: b. To assess compliance with planned arrangements and the effectiveness of the QMS.
Explanation: Regular quality audits are conducted to ensure that the implemented QMS is effective and is being complied with, ensuring continuous improvement in quality management processes.

242. When executing a large commercial building project, a contractor must select a suitable concrete mix to ensure structural integrity. Which factor is LEAST relevant in this selection?
a. The price of the concrete mix
b. The compressive strength of the concrete mix
c. The workability of the concrete mix
d. The aesthetic appeal of the finished concrete surface

Answer: d. The aesthetic appeal of the finished concrete surface
Explanation: While aesthetics are important, the structural integrity and functionality of the concrete are paramount, prioritizing compressive strength and workability over aesthetic appeal in selection.

243. A contractor has noted discrepancies between the project's quality standards and the actual quality of completed work. What is the MOST appropriate next step?
a. Ignore the discrepancies as minor issues
b. Implement immediate corrective actions to resolve the discrepancies
c. Inform the subcontractor about the discrepancies and request an explanation
d. Document the discrepancies for future reference

Answer: b. Implement immediate corrective actions to resolve the discrepancies
Explanation: Identifying discrepancies demands immediate corrective action to align the actual work quality with the project's quality standards, preventing any further complications or quality degradation.

244. When evaluating the effectiveness of implemented quality control measures in a construction project, which of the following indicators is the MOST crucial?
a. Client satisfaction
b. Number of non-conformances identified
c. Speed of construction
d. Budget adherence

Answer: a. Client satisfaction
Explanation: Ultimately, effective quality control measures should result in a final product that meets or exceeds the client's expectations, making client satisfaction a crucial indicator of effectiveness.

245. In a scenario where a contractor needs to perform a slump test for freshly mixed concrete, what is the PRIMARY purpose of this test?
a. To measure the concrete's compressive strength
b. To assess the concrete's workability
c. To determine the concrete's curing time
d. To calculate the concrete's weight

Answer: b. To assess the concrete's workability
Explanation: A slump test is conducted to measure the workability or consistency of the concrete mix, ensuring it is suitable for the job and meets quality standards.

246. When establishing a quality assurance plan, what is the MAIN focus of defining clear quality objectives?
a. To assign responsibilities and authorities
b. To enhance team morale
c. To set measurable goals and expectations to achieve the desired quality
d. To determine the budget and timeline of the project

Answer: c. To set measurable goals and expectations to achieve the desired quality. Explanation: Clear quality objectives provide a framework for what is expected in terms of quality, enabling the assessment of whether the project outputs meet the predefined standards and client requirements.

247. For a large infrastructure project, the contractor needs to ensure that all welds meet the specified quality standards. Which non-destructive testing method is MOST suitable for detecting internal defects in welds?
a. Visual inspection
b. Liquid penetrant testing
c. Ultrasonic testing
d. Magnetic particle testing

Answer: c. Ultrasonic testing. Explanation: Ultrasonic testing is highly effective in detecting internal flaws, voids, or cracks within welds, ensuring the integrity and compliance of welding work with quality standards.

248. In a complex residential project, the contractor is implementing a robust quality control system to avoid rework. What is a KEY factor in minimizing rework related to workmanship errors?
a. Strictly adhering to the project schedule
b. Enhancing communication and providing clear instructions to the workforce
c. Keeping the client informed about the progress
d. Using high-quality materials

Answer: b. Enhancing communication and providing clear instructions to the workforce
Explanation: Clear communication and instructions help in ensuring that the workforce understands the expectations and standards, thus minimizing errors and subsequent rework related to workmanship.

249. When a contractor is assessing various suppliers for procuring construction materials, which criterion is CRITICAL in ensuring the reliability and quality of the materials?
a. The location of the supplier
b. The reputation and track record of the supplier
c. The supplier's pricing structure
d. The size of the supplier's company

Answer: b. The reputation and track record of the supplier
Explanation: A supplier's reputation and track record are crucial indicators of reliability and quality, ensuring that the procured materials meet the required standards and specifications.

250. In a commercial construction project, a contractor needs to ensure that the HVAC system installation meets the required quality standards. What document is ESSENTIAL for verifying compliance with these standards?
a. Project schedule
b. Cost estimate
c. HVAC system manual
d. Quality control checklist

Answer: d. Quality control checklist
Explanation: A quality control checklist, based on the project's quality standards and specifications, is an essential tool to verify the compliance and quality of the HVAC system installation, ensuring each aspect of installation is inspected and meets the predefined criteria.

251. When integrating HVAC systems in a commercial building, which factor is paramount to ensure optimum energy efficiency and occupant comfort?
a. Size and aesthetics of the HVAC units
b. Proximity of HVAC units to electrical panels
c. Correct zoning and load calculations
d. Noise level of HVAC units

Answer: c. Correct zoning and load calculations
Explanation: Correct zoning and accurate load calculations are crucial for optimizing energy efficiency and ensuring the comfort of the occupants by providing the right amount of heating, cooling, and ventilation where needed.

252. While designing plumbing systems for a high-rise building, what consideration is crucial for maintaining water pressure in the upper floors?
a. Pipe diameter
b. Pipe material
c. Water tank elevation
d. Pipe insulation

Answer: c. Water tank elevation
Explanation: The elevation of the water tank is critical in a high-rise building to ensure that water pressure is adequate on the upper floors, utilizing gravity to maintain consistent water flow and pressure.

253. In a complex residential project, when integrating electrical and HVAC systems, what is the PRIMARY consideration for ensuring system compatibility and safety?
a. Voltage requirements of HVAC units
b. Distance between electrical and HVAC units
c. Aesthetic integration of HVAC and electrical outlets
d. Number of HVAC units per floor

Answer: a. Voltage requirements of HVAC units
Explanation: Understanding the voltage requirements of HVAC units is crucial to ensure compatibility with the electrical system and to avoid overloading, ensuring the safety and functionality of both systems.

254. For a contractor, what is the MAIN advantage of integrating Building Management Systems (BMS) for controlling HVAC, lighting, and security systems in a commercial building?
a. Reducing construction costs
b. Enhancing building aesthetics
c. Optimizing energy consumption and improving building efficiency
d. Simplifying the construction process

Answer: c. Optimizing energy consumption and improving building efficiency
Explanation: A BMS allows for the centralized control and monitoring of the building's systems, optimizing energy use, reducing operational costs, and improving overall building efficiency and performance.

255. In a scenario where a contractor is retrofitting an old building with modern plumbing systems, what is the KEY consideration to ensure proper integration with existing structures?
a. Selecting modern design fixtures
b. Ensuring minimal disruption to building occupants
c. Compatibility with existing plumbing layouts and connections
d. Selecting cost-effective plumbing materials

Answer: c. Compatibility with existing plumbing layouts and connections
Explanation: When retrofitting, ensuring the new plumbing systems are compatible with existing layouts and connections is essential to avoid complications and ensure seamless integration.

256. When integrating electrical systems in a new residential construction project, which step is crucial to prevent electrical overloads and ensure safety?
a. Installing adequate lighting fixtures
b. Installing multiple electrical panels
c. Calculating accurate electrical loads and properly sizing circuits
d. Ensuring aesthetic placement of electrical outlets

Answer: c. Calculating accurate electrical loads and properly sizing circuits
Explanation: Accurately calculating the electrical loads and properly sizing the circuits is fundamental to prevent overloads, ensuring the safety and reliability of the electrical system.

257. During the construction of a multi-story commercial building, what is the SIGNIFICANT advantage of integrating Prefabricated Vertical Building Systems in HVAC installations?
a. Reduced material costs
b. Enhanced aesthetic appeal
c. Accelerated construction timelines
d. Increased structural integrity

Answer: c. Accelerated construction timelines
Explanation: Prefabricated Vertical Building Systems allow for quicker installations of HVAC systems due to pre-assembled units, reducing on-site labor and accelerating overall construction timelines.

258. In the integration of plumbing systems in a commercial building, why is it VITAL to ensure proper slope in waste pipes?
a. To increase water flow speed
b. To prevent water hammer
c. To avoid blockages and ensure smooth flow of wastewater
d. To reduce water consumption

Answer: c. To avoid blockages and ensure smooth flow of wastewater
Explanation: Proper slope in waste pipes is essential to allow gravity to facilitate the smooth flow of wastewater, preventing blockages and maintaining the functionality of the plumbing system.

259. For a contractor integrating electrical systems in a commercial project, what is the PRIMARY consideration when selecting the appropriate wire size?
a. The cost of the wire
b. The length of the wire run
c. The color of the insulation
d. The current-carrying capacity of the wire

Answer: d. The current-carrying capacity of the wire
Explanation: The wire size must be chosen based on its ability to safely carry the current it will serve, preventing overheating and potential fire hazards, ensuring safety and compliance with electrical codes.

260. In a scenario where a contractor is integrating an HVAC system in a renovated commercial building, what is a CRUCIAL consideration to maintain indoor air quality and occupant comfort?
a. Aesthetic placement of vents
b. The noise level of the HVAC units
c. Proper sizing and selection of HVAC equipment
d. The brand reputation of the HVAC units

Answer: c. Proper sizing and selection of HVAC equipment
Explanation: Correct sizing and selection of HVAC equipment are vital to maintain the desired indoor temperature, humidity levels, and air quality, ensuring occupant comfort and well-being.

261. While conducting a land survey for a commercial building project, which method is imperative for gathering detailed and accurate topographical data?
a. Aerial survey
b. Chain survey
c. Plane table survey
d. Total station survey

Answer: d. Total station survey
Explanation: Total station surveys are crucial for acquiring precise and detailed topographical data due to their ability to measure angles and distances accurately, allowing for exact plotting of the site features and elevations.

262. In preparing a site for construction in an urban environment, what is a PRIMARY consideration to ensure the structural stability of adjacent buildings?
a. Implementing sound barriers
b. Adequate shoring and underpinning
c. Construction speed
d. Aesthetic integration with surrounding structures

Answer: b. Adequate shoring and underpinning
Explanation: Adequate shoring and underpinning are critical to support adjacent structures, preventing any potential ground movement or subsidence that could compromise their structural stability during construction.

263. When selecting a reference line for a construction project, which factor is ESSENTIAL to ensure the accurate layout of the structure?
a. Proximity to the construction site
b. Visibility from the main road
c. Perpendicularity or parallelism to the main structure
d. Distance from adjacent buildings

Answer: c. Perpendicularity or parallelism to the main structure
Explanation: A reference line must be either perpendicular or parallel to the main structure to serve as an accurate baseline for laying out the building components and ensuring the correct alignment of the structure.

264. During site preparation for a commercial building project, what is the SIGNIFICANT reason for conducting a thorough soil analysis?
a. To determine the landscaping possibilities
b. To calculate the cost of soil removal
c. To assess soil's load-bearing capacity and determine foundation design
d. To estimate the volume of soil to be excavated

Answer: c. To assess soil's load-bearing capacity and determine foundation design
Explanation: Thorough soil analysis is crucial to assess the soil's load-bearing capacity, which directly influences the design of the building's foundation, ensuring the stability and safety of the structure.

265. When conducting a land survey for a large residential development, what is the PRIMARY purpose of establishing control points across the site?
a. To mark the boundaries of the property
b. To provide reference points for accurate measurements and alignments
c. To indicate the locations of future structures
d. To designate areas for landscaping

Answer: b. To provide reference points for accurate measurements and alignments
Explanation: Establishing control points is vital to provide accurate reference points across the site for measurements and alignments, ensuring precise layout and construction of the development.

266. In a scenario where a contractor is preparing a sloped site for construction, which technique is crucial to prevent soil erosion and stabilize the slope?
a. Installing sound barriers
b. Applying hydroseeding
c. Implementing surface drainage systems
d. Erecting temporary fences

Answer: c. Implementing surface drainage systems
Explanation: Surface drainage systems are essential on sloped sites to control the flow of water, prevent soil erosion, and stabilize the slope, ensuring the integrity of the site and construction project.

267. During a construction project, why is it ESSENTIAL to consider the legal and physical constraints of a site during land surveying and site preparation phases?
a. To facilitate construction equipment placement
b. To avoid disputes with adjacent property owners and ensure compliance with local regulations
c. To determine the aesthetic aspects of the project
d. To estimate construction costs accurately

Answer: b. To avoid disputes with adjacent property owners and ensure compliance with local regulations
Explanation: Considering legal and physical constraints is crucial to avoid any legal disputes with adjacent property owners and ensure that all aspects of land surveying and site preparation comply with local zoning and building regulations.

268. When preparing a site for the construction of a high-rise building, which type of foundation is MOST suitable to distribute the load to the subsoil efficiently?
a. Strip foundation
b. Raft foundation
c. Pile foundation
d. Pad foundation

Answer: c. Pile foundation
Explanation: For high-rise buildings, a pile foundation is most suitable as it can transfer heavy loads from the structure through weaker soils to stronger, more stable soils or rock at greater depths, ensuring the stability of the building.

269. For a contractor working on a mixed-use development project, why is the accurate identification and marking of underground utilities CRITICAL during site preparation?
a. To facilitate the landscaping process
b. To prevent disruptions, damages, and ensure the safety of the construction site
c. To determine the construction timeline
d. To assess the aesthetic impact of the utilities on the project

Answer: b. To prevent disruptions, damages, and ensure the safety of the construction site
Explanation: Accurately identifying and marking underground utilities is crucial to avoid damaging them during excavation, preventing potential disruptions, costly repairs, and ensuring the safety of workers and the public.

270. In a scenario where a land surveyor is using GPS technology for surveying a rural construction site, what is the MAIN advantage of utilizing Real-Time Kinematic (RTK) GPS?
a. Extended signal range
b. Increased battery life
c. Enhanced signal stability
d. Improved positional accuracy

Answer: d. Improved positional accuracy
Explanation: Real-Time Kinematic (RTK) GPS provides centimeter-level accuracy in real-time, making it highly beneficial for land surveying where precise measurements and positional accuracy are paramount.

271. A construction project requires the excavation of a rectangular area measuring 30m x 20m x 5m. If the soil has a swell factor of 20%, how much total volume will the excavated soil occupy?
a. 3,600 m³
b. 4,320 m³
c. 6,000 m³
d. 7,200 m³

Answer: b. 4,320 m³
Explanation: The original volume is found using the formula for the volume of a rectangular prism: Length x Width x Height. 30m x 20m x 5m = 3,000 m³. With a 20% swell factor, the excavated soil will occupy 3,000 m³ x 1.20 = 3,600 m³ additional volume, totaling 4,320 m³.

Answer: b. It represents the beam's resistance to bending.
Explanation: In the Flexural Formula, the Moment of Inertia (I) represents the beam's ability to resist bending. It quantifies how the cross-sectional area of a beam is distributed about the neutral axis, impacting its flexural strength.

272. For a concrete slab with a compressive strength of 3,000 psi, what would be the safe load capacity per square foot if the factor of safety is 4?
a. 750 psf
b. 1,000 psf
c. 1,250 psf
d. 1,500 psf

Answer: a. 750 psf

273. An architect designs a trapezoidal garden with parallel sides measuring 15m and 25m, and a height of 10m. What is the area of the garden?
a. 200 m²
b. 250 m²
c. 300 m²
d. 350 m²

Answer: a. 200 m²

274. If a project's budget is $500,000 and the contractor's mark-up is 15%, what will be the total bid price submitted to the client?
a. $525,000
b. $550,000
c. $575,000
d. $575,500

Answer: c. $575,000

275. The center-to-center spacing of studs in a wall framing is 16 inches. If the wall length is 32 feet, how many studs are required?
a. 24
b. 25
c. 26
d. 27

Answer: a. 24. Explanation: First, convert the wall length to inches: 32 feet x 12 inches/foot = 384 inches. Then, divide by the spacing: 384 inches / 16 inches/stud = 24 studs.

276. A cylindrical water tank has a diameter of 3m and a height of 5m. How many cubic meters of water can it hold?
a. 35.32 m³
b. 44.21 m³
c. 47.12 m³
d. 52.35 m³

Answer: a. 35.32 m³

277. A project has a total of 250 man-days to complete. If a worker works 8 hours a day, how many total man-hours are required to complete the project?
a. 2,000 man-hours
b. 2,250 man-hours
c. 2,500 man-hours
d. 2,750 man-hours

Answer: a. 2,000 man-hours
Explanation: To convert man-days to man-hours, multiply by the number of work hours per day: 250 man-days * 8 hours/day = 2,000 man-hours.

278. During the calculation of a project's earthwork, the contractor must consider a soil shrinkage factor of 10%. If the project requires 10,000 cubic yards of compacted soil, how many cubic yards of loose soil should be acquired?
a. 9,000 cu yd
b. 10,000 cu yd
c. 11,000 cu yd
d. 11,100 cu yd

Answer: c. 11,000 cu yd

279. A contractor needs to calculate the thermal resistance (R-value) of a wall. The wall is constructed with a material having a thermal conductivity (k-value) of 0.5 W/m·K and a thickness of 0.2m. What is the R-value of the wall?
a. 0.1 m²·K/W
b. 0.4 m²·K/W
c. 2.5 m²·K/W
d. 4.0 m²·K/W

Answer: c. 2.5 m²·K/W

280. In a green construction project, what is the primary role of a building envelope with high thermal mass?
a. Enhance indoor air quality.
b. Minimize solar heat gain.
c. Optimize daylighting.
d. Regulate indoor temperatures.

Answer: d. Regulate indoor temperatures.
Explanation: High thermal mass in a building envelope absorbs and stores heat, helping to stabilize indoor temperatures by reducing temperature fluctuations, thereby enhancing energy efficiency and comfort.

281. When employing passive solar design principles, the orientation of the building should primarily aim to:
a. Maximize wind exposure.
b. Maximize summer solar gain.
c. Maximize winter solar gain.
d. Minimize daylighting.

Answer: c. Maximize winter solar gain.
Explanation: Passive solar design aims to maximize winter solar gain to reduce heating needs while minimizing it during summer to avoid overheating, thus optimizing energy efficiency.

282. What is the primary advantage of utilizing a green roof in urban construction projects?
a. Enhanced aesthetic appeal.
b. Increased structural loading.
c. Reduction of heat island effect.
d. Enhanced photovoltaic efficiency.

Answer: c. Reduction of heat island effect.
Explanation: Green roofs help mitigate the urban heat island effect by absorbing less heat and reflecting more solar radiation than traditional roofing materials, thereby reducing ambient temperatures.

283. In LEED-certified buildings, which water conservation strategy can be considered the most efficient for reducing water consumption in landscapes?
a. Utilization of graywater.
b. Implementation of high-efficiency irrigation systems.
c. Use of drought-tolerant plant species.
d. Reduction of vegetated areas.

Answer: a. Utilization of graywater.
Explanation: Utilizing graywater for landscape irrigation is a highly effective way to conserve water, as it reduces the demand for freshwater and efficiently recycles previously used water.

284. The implementation of Building Management Systems (BMS) primarily aids in:
a. Monitoring and controlling indoor environmental parameters.
b. Enhancing the structural integrity of the building.
c. Improving the aesthetic appeal of the building.
d. Reducing construction cost.

Answer: a. Monitoring and controlling indoor environmental parameters.
Explanation: Building Management Systems are crucial for optimizing building performance by continuously monitoring and controlling indoor environmental parameters such as temperature, humidity, and lighting, thus aiding in energy conservation and improving occupant comfort.

285. When considering life-cycle assessment in sustainable construction, the "Cradle to Grave" approach involves:
a. Analyzing only the operational phase of a building.
b. Assessing only the construction and design phases.
c. Evaluating the entire life cycle, from raw material extraction to disposal.
d. Focusing only on the demolition and waste management phase.

Answer: c. Evaluating the entire life cycle, from raw material extraction to disposal.
Explanation: A "Cradle to Grave" life-cycle assessment evaluates every phase of a product's life cycle, from raw material extraction through materials processing, manufacturing, distribution, use, repair and maintenance, and disposal or recycling.

286. For a construction project located in a water-scarce region, which sustainable approach would be the most effective in ensuring water conservation?
a. Installing high-efficiency HVAC systems.
b. Employing rainwater harvesting systems.
c. Incorporating high albedo materials.
d. Maximizing natural ventilation.

Answer: b. Employing rainwater harvesting systems.
Explanation: In water-scarce regions, employing rainwater harvesting systems is crucial as it allows the collection and use of rainwater, reducing reliance on conventional water sources and promoting water conservation.

287. In a Net Zero Energy Building, the energy consumption is primarily offset by:
a. Purchasing Renewable Energy Certificates (RECs).
b. Implementing energy-efficient lighting and HVAC systems.
c. Employing passive design strategies.
d. Generating renewable energy on-site.

Answer: d. Generating renewable energy on-site.
Explanation: A Net Zero Energy Building generates as much renewable energy on-site as it consumes on an annual basis, primarily through solar panels, wind turbines, or other renewable energy systems, effectively balancing out its energy consumption.

288. What is the primary reason for implementing cool roof technology in sustainable buildings located in hot climates?
a. To allow rainwater harvesting.
b. To minimize solar heat gain.
c. To increase roof durability.
d. To maximize photovoltaic efficiency.

Answer: b. To minimize solar heat gain.
Explanation: Cool roof technology reflects more sunlight and absorbs less heat than standard roofs, minimizing solar heat gain, reducing cooling loads, and contributing to thermal comfort in buildings located in hot climates.

289. In sustainable construction, what is the primary goal of utilizing low VOC (Volatile Organic Compounds) materials?
a. To increase energy efficiency.
b. To improve indoor air quality.
c. To enhance thermal comfort.
d. To minimize construction cost.

Answer: b. To improve indoor air quality.
Explanation: Low VOC materials release fewer pollutants into the air, reducing the risk of indoor air contamination and ensuring healthier and more comfortable living and working environments.

290. When designing a building to be energy efficient, which strategy would be most effective in reducing cooling loads in a hot and sunny climate?
a. Utilizing high thermal mass materials.
b. Implementing extensive glazing on the west façade.
c. Employing high albedo roofing materials.
d. Increasing insulation on the interior walls.

Answer: c. Employing high albedo roofing materials.
Explanation: High albedo roofing materials reflect more sunlight, minimizing solar heat gain and reducing cooling loads in hot and sunny climates, thus contributing to energy efficiency.

291. In a scenario where a construction project aims for optimum energy conservation, what is the primary purpose of integrating a ground-source heat pump system?
a. To provide onsite electricity generation.
b. To facilitate stormwater management.
c. To utilize the constant temperature of the ground for heating and cooling.
d. To recover heat from the building's wastewater.

Answer: c. To utilize the constant temperature of the ground for heating and cooling.
Explanation: Ground-source heat pump systems exploit the earth's consistent temperatures to provide heating in the winter and cooling in the summer, thus optimizing energy conservation.

292. When selecting sustainable materials for a construction project, a contractor would prioritize materials that have:
a. High embodied energy and are non-renewable.
b. Low embodied energy and are renewable.
c. High thermal conductivity and high density.
d. Low albedo and high emissivity.

Answer: b. Low embodied energy and are renewable.
Explanation: Prioritizing materials with low embodied energy and renewable properties is crucial for sustainability as they have less environmental impact in terms of energy consumption and resource depletion.

293. A construction project located in a region with frequent power outages incorporates a Photovoltaic (PV) system with battery storage. What is the primary advantage of this combination?
a. Enhanced PV efficiency.
b. Increased energy generation.
c. Continuous power supply during outages.
d. Reduction in PV system installation cost.

Answer: c. Continuous power supply during outages.
Explanation: Integrating battery storage with a PV system allows for the storage of excess generated energy, ensuring a continuous power supply during power outages, enhancing resilience and reliability.

294. What is the primary consideration when integrating wind turbines into a building design for energy generation?
a. Proximity to bodies of water.
b. Amount of daily sunlight.
c. Local wind speed and direction.
d. Proximity to urban areas.

Answer: c. Local wind speed and direction.
Explanation: The effectiveness of wind turbines is highly dependent on local wind conditions; therefore, understanding local wind speed and direction is critical to optimize energy generation.

295. When employing passive design strategies for an office building located in a temperate climate, which technique would primarily enhance daylighting?
a. Incorporating overhangs and shading devices.
b. Using thermal mass and insulation.
c. Designing open floor plans and utilizing reflective interior finishes.
d. Installing high-efficiency HVAC systems.

Answer: c. Designing open floor plans and utilizing reflective interior finishes.
Explanation: Open floor plans and reflective interior finishes maximize the penetration and reflection of natural light within the building, enhancing daylighting and reducing the need for artificial lighting.

296. What is the primary advantage of specifying Fly Ash in concrete mixes for sustainable construction projects?
a. To increase the water content in the mix.
b. To reduce the carbon footprint associated with cement production.
c. To enhance the aesthetic appeal of the concrete.
d. To decrease the curing time of the concrete.

Answer: b. To reduce the carbon footprint associated with cement production.
Explanation: Utilizing fly ash in concrete mixes substitutes for cement, which is energy-intensive to produce, thus reducing the carbon footprint and enhancing the sustainability of the project.

297. In an off-grid construction project, which renewable energy system would be most suitable for a location with limited sunlight and consistent high winds?
a. Solar Thermal Systems.
b. Photovoltaic Systems.
c. Wind Turbine Systems.
d. Biomass Energy Systems.

Answer: c. Wind Turbine Systems.
Explanation: In locations with limited sunlight but consistent high winds, wind turbine systems are more suitable for energy generation, providing a reliable source of renewable energy.

298. A contractor aiming for enhanced energy efficiency in a residential building should prioritize the installation of:
a. Low-E windows.
b. Incandescent light bulbs.
c. Standard-efficiency appliances.
d. Single-glazed windows.

Answer: a. Low-E windows.
Explanation: Low-E windows minimize the amount of infrared and ultraviolet light that can pass through glass without compromising the amount of visible light, thus contributing to energy efficiency by reducing heat gain or loss.

299. When considering sustainable material selection for a new construction project, the decision to utilize reclaimed wood primarily aims to:
a. Reduce cost associated with new wood procurement.
b. Minimize the need for structural reinforcement.
c. Reduce demand on new wood resources and decrease waste in landfills.
d. Improve the aesthetic appeal of the building.

Answer: c. Reduce demand on new wood resources and decrease waste in landfills. Explanation: Utilizing reclaimed wood is sustainable as it reduces the pressure on forests and ecosystems for new wood resources and minimizes waste by reusing materials that would otherwise end up in landfills.

300. To optimize water conservation, a contractor decides to implement a rainwater harvesting system for a new commercial building. What is the primary benefit of this system?
a. Reduction in construction costs.
b. Reduction in the building's energy consumption.
c. Provision of an alternate water source for non-potable uses.
d. Enhancement of the building's aesthetic appeal.

Answer: c. Provision of an alternate water source for non-potable uses.
Explanation: Rainwater harvesting systems collect and store rainwater, providing an alternate source of water primarily for non-potable uses, such as irrigation, thereby conserving potable water resources.

301. When prioritizing sustainable site development, what is the main objective of incorporating permeable paving materials?
a. To increase onsite parking availability.
b. To reduce heat island effect.
c. To facilitate stormwater management.
d. To enhance the visual appeal of the site.

Answer: c. To facilitate stormwater management.
Explanation: Permeable paving materials allow water to permeate through the surface to the underlying layers, reducing surface runoff and facilitating stormwater management.

302. In the context of indoor environmental quality, what is the primary goal of utilizing low-VOC materials within a building?
a. To reduce material costs.
b. To enhance material durability.
c. To improve indoor air quality.
d. To increase energy efficiency.

Answer: c. To improve indoor air quality.
Explanation: Low-VOC materials release fewer volatile organic compounds, contributing to better indoor air quality and reducing health risks associated with poor air quality.

303. To advance waste management and recycling during a construction project, what is the most sustainable approach to dealing with construction waste?
a. Disposing of all waste in landfills.
b. Implementing waste-to-energy processes for all waste materials.
c. Maximizing waste segregation and recycling.
d. Opting for single-stream recycling for all waste materials.

Answer: c. Maximizing waste segregation and recycling.
Explanation: Maximizing waste segregation and recycling ensures that materials are processed and reused efficiently, minimizing landfill disposal and resource depletion.

304. For a residential development project aiming for sustainability, what is the primary consideration when selecting plant species for landscaping?
a. Aesthetic appeal of the species.
b. Size of the plant species at maturity.
c. Availability of the species in local nurseries.
d. Adaptation of the species to local climate and conditions.

Answer: d. Adaptation of the species to local climate and conditions.
Explanation: Selecting plant species that are well-adapted to local conditions ensures their survival and growth with minimal resource input, contributing to sustainable landscaping practices.

305. What is the main purpose of implementing greywater systems in buildings aiming for water conservation and sustainability?
a. To reduce water heating costs.
b. To provide an alternative source of potable water.
c. To reuse wastewater for non-potable applications.
d. To minimize the need for water treatment.

Answer: c. To reuse wastewater for non-potable applications.
Explanation: Greywater systems collect and treat wastewater from sources like sinks and showers, allowing its reuse for non-potable applications such as irrigation and toilet flushing, thus conserving water resources.

306. A contractor decides to implement a green roof on a new commercial building. What is the primary environmental benefit of this implementation?
a. Reduction in building's energy consumption.
b. Enhancement of building's aesthetic appeal.
c. Increase in the building's resale value.
d. Provision of additional commercial space.

Answer: a. Reduction in building's energy consumption.
Explanation: Green roofs provide insulation and reduce heat island effect, thus reducing the energy required for heating and cooling, which contributes to overall energy conservation.

307. When developing a construction waste management plan, what is the fundamental principle to optimize waste diversion from landfills?
a. Maximizing waste incineration.
b. Maximizing waste compaction.
c. Maximizing waste reduction and recycling.
d. Maximizing waste transportation efficiency.

Answer: c. Maximizing waste reduction and recycling.
Explanation: To optimize waste diversion from landfills, it is essential to prioritize waste reduction, reuse, and recycling, which helps conserve resources and reduce environmental impacts.

308. In sustainable site development, what is the primary environmental advantage of utilizing native plant species in landscaping?
a. They require minimal maintenance and resources compared to non-native species.
b. They have a higher aesthetic appeal compared to non-native species.
c. They have a higher resistance to pests compared to non-native species.
d. They have a faster growth rate compared to non-native species.

Answer: a. They require minimal maintenance and resources compared to non-native species.
Explanation: Native plant species are adapted to local conditions, thus requiring less water, fertilizer, and maintenance compared to non-native species, contributing to environmental sustainability.

309. For a construction project looking to enhance indoor environmental quality, what is the key benefit of integrating a demand-controlled ventilation system?
a. Reduction in building's water consumption.
b. Improvement in indoor air quality through optimized ventilation.
c. Enhancement of building's aesthetic appeal.
d. Increase in building's resale value.

Answer: b. Improvement in indoor air quality through optimized ventilation.
Explanation: Demand-controlled ventilation systems adjust the ventilation rate in real-time based on occupancy and air quality needs, ensuring optimal indoor air quality and reducing energy consumption.

310. When establishing a safety perimeter around a construction site, what is the primary consideration for the designated safety officer?
a. Aesthetic alignment of the safety barriers.
b. Distance from the public right-of-way.
c. Cost of setting up the safety perimeter.
d. Proximity to potential hazards within the construction site.

Answer: d. Proximity to potential hazards within the construction site.
Explanation: The primary consideration is the proximity to potential hazards; the safety perimeter should be established at a safe distance to protect both workers and the public from construction site hazards.

311. In the event of a fire on a construction site, what is the initial action to be undertaken by the site supervisor present at the scene?
a. Attempt to extinguish the fire using available resources.
b. Evacuate all personnel and then call the emergency services.
c. Secure the construction equipment and materials.
d. Assess the damage caused by the fire.

Answer: b. Evacuate all personnel and then call the emergency services.
Explanation: The initial action must be to ensure the safety of all personnel by evacuating them from danger and then promptly call the emergency services to address the fire.

312. When selecting appropriate fall protection systems for workers operating above ground level, what factor is critical?
a. The cost of the fall protection system.
b. The comfort of the fall protection system for the worker.
c. The height at which the worker is operating.
d. The mobility of the worker when using the fall protection system.

Answer: c. The height at which the worker is operating.
Explanation: The height at which the worker is operating is crucial in determining the type and extent of fall protection required to mitigate the risk of fall-related injuries effectively.

313. In a scenario where workers are exposed to airborne contaminants, what is the primary responsibility of the contractor?
a. To provide suitable respiratory protection to the workers.
b. To conduct regular air quality tests.
c. To limit the work hours of exposed workers.
d. To educate workers on the effects of airborne contaminants.

Answer: a. To provide suitable respiratory protection to the workers.
Explanation: The contractor must ensure the safety of the workers by providing suitable respiratory protection to minimize the inhalation of harmful airborne contaminants.

314. To ensure electrical safety on a construction site, what is the essential safety measure to implement when using electrical equipment?
a. Use of battery-operated equipment.
b. Regular inspection and maintenance of electrical equipment.
c. Use of equipment with the highest available voltage.
d. Use of equipment with the longest cords.

Answer: b. Regular inspection and maintenance of electrical equipment.
Explanation: Regular inspection and maintenance of electrical equipment are crucial to prevent electrical hazards, such as shocks and fires, ensuring the safety of workers on site.

315. For managing noise hazards on a construction site, what is the most effective control measure to implement initially?
a. Providing ear protection to all workers.
b. Erecting noise barriers around the site.
c. Scheduling noisy work during low occupancy times.
d. Employing quieter construction methods and equipment.

Answer: d. Employing quieter construction methods and equipment.
Explanation: The most effective initial control measure is to employ quieter construction methods and equipment to reduce the noise levels at the source, mitigating the risk to workers and the surrounding community.

316. When constructing a trench, what is the primary consideration to ensure worker safety during the excavation process?
a. The speed of the excavation process.
b. The proximity of the excavation to existing structures.
c. The cost of excavation equipment.
d. The soil type and condition.

Answer: d. The soil type and condition.
Explanation: Understanding the soil type and condition is critical to determine the stability of the trench and implement appropriate protective systems to prevent collapses and ensure worker safety.

317. To maintain a safe working environment, what is the crucial safety protocol regarding the storage of construction materials on site?
a. Store materials in close proximity to the work area.
b. Store materials based on the order of use.
c. Store materials securely to prevent unauthorized access.
d. Store materials in a manner that minimizes manual handling.

Answer: d. Store materials in a manner that minimizes manual handling.
Explanation: Proper storage that minimizes manual handling is essential to prevent strains, sprains, and other injuries associated with moving and handling construction materials.

318. In the event of a chemical spill on a construction site, what is the immediate action required by site personnel?
a. Attempt to contain the spill using available resources.
b. Evacuate the area and notify the designated safety officer.
c. Identify the chemical and research cleanup procedures.
d. Continue work but avoid the contaminated area.

Answer: b. Evacuate the area and notify the designated safety officer.
Explanation: The immediate action is to evacuate the affected area to ensure the safety of site personnel and then notify the designated safety officer to manage the response and cleanup process appropriately.

319. When implementing a traffic management plan around a construction site, what is the fundamental safety objective?
a. To minimize disruption to local traffic.
b. To ensure smooth flow of construction vehicles.
c. To protect construction workers and the general public from traffic-related risks.
d. To reduce the cost of traffic management implementation.

Answer: c. To protect construction workers and the general public from traffic-related risks.
Explanation: The fundamental objective is to manage traffic effectively to protect both construction workers and the general public from any potential traffic-related risks around the construction site.

320. When selecting PPE for workers using cutting tools on a construction site, what factor is paramount?
a. Affordability of the PPE.
b. Comfort and fit for the worker.
c. Durability and lifespan of the PPE.
d. Level of protection offered by the PPE.

Answer: d. Level of protection offered by the PPE.
Explanation: When selecting PPE, the paramount consideration should be the level of protection offered to the user against specific hazards associated with the use of cutting tools.

321. In the context of fall protection, what characteristic is critical for the full-body harness used as PPE in elevated work areas?
a. Lightweight design.
b. Elasticity of the material.
c. Adequate load-bearing capacity.
d. Color visibility.

Answer: c. Adequate load-bearing capacity.
Explanation: An adequate load-bearing capacity is critical to ensure the harness can effectively support the worker's weight and arrest a fall, preventing injuries or fatalities.

322. For workers involved in the handling of hazardous chemicals, what type of PPE is essential?
a. High-visibility clothing.
b. Chemical-resistant gloves.
c. Acoustic earmuffs.
d. Steel-toed boots.

Answer: b. Chemical-resistant gloves.
Explanation: Chemical-resistant gloves are essential to protect the hands from chemical exposure, preventing skin irritations, burns, and other injuries.

323. When implementing respiratory protection for workers exposed to airborne particulates, what factor should be prioritized in the selection of respirators?
a. The ease of communication while wearing the respirator.
b. The filtration efficiency for the specific particulate.
c. The cost per unit of the respirator.
d. The versatility of use with other PPE.

Answer: b. The filtration efficiency for the specific particulate.
Explanation: Filtration efficiency should be prioritized to ensure that the respirator can effectively filter out the specific airborne particulates to which the workers are exposed.

324. In a scenario where eye hazards are present, what is the most suitable type of eye protection?
a. Standard safety glasses.
b. Direct-vented goggles.
c. Indirect-vented goggles.
d. Face shield.

Answer: c. Indirect-vented goggles.
Explanation: Indirect-vented goggles offer protection against splashes, droplets, and dust, making them suitable for environments with multiple eye hazards.

325. For construction work in a noisy environment, which type of hearing protection is most appropriate for intermittent noise exposure?
a. Canal caps.
b. Earmuffs.
c. Custom molded earplugs.
d. Foam earplugs.

Answer: a. Canal caps.
Explanation: Canal caps are suitable for environments with intermittent noise as they can be easily put on and taken off, providing convenience and protection for workers.

326. When working in an environment with falling objects, what characteristic of head protection is crucial?
a. The color and visibility of the hard hat.
b. The ventilation provided by the hard hat.
c. The shock absorption capacity of the hard hat.
d. The weight of the hard hat.

Answer: c. The shock absorption capacity of the hard hat.
Explanation: The ability of the hard hat to absorb shock is critical to protect the wearer's head from the impact of falling objects.

327. In a construction scenario where workers are exposed to welding operations, what PPE is essential to protect against ultraviolet radiation?
a. Sunscreen.
b. Welding curtains.
c. UV-blocking safety glasses.
d. Full-face respirator.

Answer: c. UV-blocking safety glasses.
Explanation: UV-blocking safety glasses are essential to protect the eyes from harmful ultraviolet radiation emitted during welding operations.

328. To ensure optimal foot protection in an area with a risk of puncture wounds, what feature should be prioritized in selecting footwear?
a. Slip-resistant soles.
b. Steel toe caps.
c. Puncture-resistant soles.
d. Insulated lining.

Answer: c. Puncture-resistant soles.
Explanation: Puncture-resistant soles should be prioritized to protect the feet from sharp objects like nails, minimizing the risk of puncture wounds.

329. When conducting a risk assessment to determine the need for PPE, what is the primary consideration?
a. Availability of PPE.
b. Cost of acquiring PPE.
c. Types of hazards present.
d. Worker preference for PPE.

Answer: c. Types of hazards present.
Explanation: Identifying the types of hazards present is the primary consideration in conducting a risk assessment for PPE, as it dictates the appropriate PPE needed to mitigate those specific risks.

330. A construction manager is assessing the need for fall protection on a site with multiple elevated work areas. What is the crucial factor in determining the requirement for fall protection systems?
a. Height of the work area from the ground.
b. Number of workers on the elevated surface.
c. Proximity to the edge of the elevated surface.
d. Duration of work on the elevated surface.

Answer: a. Height of the work area from the ground.
Explanation: The height of the work area from the ground is crucial in assessing the need for fall protection, as it determines the potential fall distance and risk of injury.

331. Which component of a personal fall arrest system is primarily responsible for reducing the impact forces during a fall?
a. Anchor point.
b. Lanyard.
c. Full-body harness.
d. Energy absorber.

Answer: d. Energy absorber.
Explanation: Energy absorbers are designed to extend and absorb energy during a fall, reducing the forces exerted on the worker's body.

332. In scenarios involving work near the edge of a roof, what fall protection measure is most effective in preventing falls?
a. Warning line system.
b. Guardrail system.
c. Safety net system.
d. Personal fall arrest system.

Answer: b. Guardrail system.
Explanation: Guardrail systems are effective barriers that prevent workers from falling off edges, providing a passive form of fall protection.

333. What is the primary purpose of conducting a fall hazard assessment prior to commencing work at height?
a. To identify the types of fall protection equipment available.
b. To determine the cost implications of implementing fall protection.
c. To identify the locations and risks of potential fall hazards.
d. To assess the worker's proficiency in using fall protection equipment.

Answer: c. To identify the locations and risks of potential fall hazards.
Explanation: A fall hazard assessment aims to identify potential fall hazards and their risks to implement appropriate fall protection measures.

334. What is the maximum allowable free fall distance when using a personal fall arrest system?
a. 4 feet.
b. 6 feet.
c. 10 feet.
d. 15 feet.

Answer: b. 6 feet.
Explanation: OSHA regulations stipulate that the maximum allowable free fall distance for a personal fall arrest system is 6 feet.

335. In construction projects involving scaffolding, at what height is fall protection required?
a. Above 6 feet.
b. Above 10 feet.
c. Above 15 feet.
d. Above 20 feet.

Answer: b. Above 10 feet.
Explanation: OSHA mandates the use of fall protection on scaffolds when workers are at heights of 10 feet or more above a lower level.

336. When implementing a controlled access zone as a fall protection measure, what is a key requirement?
a. The presence of a competent person to monitor the zone.
b. The installation of a warning line system around the zone.
c. The use of personal fall arrest systems within the zone.
d. The prohibition of work at heights within the zone.

Answer: a. The presence of a competent person to monitor the zone.
Explanation: A controlled access zone requires a competent person to monitor the zone and ensure that only authorized personnel enter and work within it.

337. When selecting anchor points for personal fall arrest systems, what characteristic is critical?
a. Accessibility.
b. Visibility.
c. Strength and stability.
d. Proximity to work area.

Answer: c. Strength and stability.
Explanation: Anchor points must have sufficient strength and stability to support the forces generated during a fall and arrest it safely.

338. Which type of fall protection system is designed to arrest a worker's fall before contact with a lower level?
a. Guardrail system.
b. Safety net system.
c. Positioning device system.
d. Personal fall arrest system.

Answer: d. Personal fall arrest system.
Explanation: Personal fall arrest systems are designed to stop a worker's fall before reaching a lower level, preventing contact and minimizing the risk of injury.

339. In a scenario where construction workers are required to perform tasks on steep roofs, what type of fall protection is appropriate?
a. Safety monitoring system.
b. Warning line system.
c. Guardrail system.
d. Personal fall arrest system.

Answer: d. Personal fall arrest system.
Explanation: On steep roofs where the risk of falling is high, personal fall arrest systems are suitable to arrest falls and prevent injuries.

340. A construction project is underway involving the use of various hazardous chemicals. What is the primary document responsible for conveying information about these chemicals?
a. Safety Plan.
b. Emergency Action Plan.
c. Hazard Communication Plan.
d. Incident Report.

Answer: c. Hazard Communication Plan.

Explanation: The Hazard Communication Plan is critical for providing information on the presence of hazardous chemicals in the workplace, outlining procedures, and educating employees on safe handling practices.

341. What is the significance of a Safety Data Sheet (SDS) in Hazard Communication?
a. Outlining emergency escape routes.
b. Providing specific information about chemical hazards and safe handling.
c. Detailing the company's safety policies and procedures.
d. Documenting workplace accidents and incidents.

Answer: b. Providing specific information about chemical hazards and safe handling.

Explanation: SDSs are vital documents that give detailed information about the properties, hazards, safe handling, and emergency measures related to chemical products.

342. When should employees be trained on the hazards of new chemicals introduced to the workplace?
a. Within one week of the chemical's introduction.
b. Before they start work with the new chemical.
c. At the next scheduled training session.
d. Within a month of the chemical's introduction.

Answer: b. Before they start work with the new chemical.

Explanation: Employees must be trained on the hazards of a new chemical before they begin working with it to ensure they understand the risks and can handle it safely.

343. What is the primary purpose of labeling in Hazard Communication?
a. To list the ingredients of the chemical.
b. To identify hazards and convey safety information quickly.
c. To provide contact information of the manufacturer.
d. To display the price of the chemical product.

Answer: b. To identify hazards and convey safety information quickly.

Explanation: Labels on hazardous chemicals serve to swiftly communicate the hazards and precautionary measures associated with the chemicals, aiding in immediate recognition and safe handling.

344. In a scenario where a contractor is managing a site with various subcontractors utilizing hazardous chemicals, who is primarily responsible for ensuring Hazard Communication standards are met?
a. Subcontractors.
b. The General Contractor.
c. The Owner of the property.
d. Individual Workers.

Answer: b. The General Contractor.
Explanation: The General Contractor bears the overall responsibility for compliance with Hazard Communication standards, ensuring that all subcontractors and workers are informed of and adhere to safe practices.

345. Which section of the Safety Data Sheet (SDS) provides information about the safe storage conditions for a chemical?
a. Section 7: Handling and Storage.
b. Section 2: Hazards Identification.
c. Section 4: First-Aid Measures.
d. Section 8: Exposure Controls/Personal Protection.

Answer: a. Section 7: Handling and Storage.
Explanation: Section 7 of the SDS provides explicit information regarding the safe handling and storage conditions of a chemical to prevent accidents and exposure.

346. The construction manager noticed unlabeled containers of chemicals on site. What immediate action is required?
a. Reporting to OSHA.
b. Relocating the containers to a secure area.
c. Ensuring the containers are labeled properly.
d. Disposing of the containers.

Answer: c. Ensuring the containers are labeled properly.
Explanation: Unlabeled containers pose a risk, and it is crucial to immediately label them correctly with the necessary hazard and precautionary information to ensure safety.

347. A worker noticed the absence of an SDS for a chemical recently delivered to the site. What should be his first course of action?
a. Continue working but avoid contact with the chemical.
b. Request the SDS from the supplier or manufacturer.
c. Notify the site safety officer immediately.
d. Return the chemical to the supplier.

Answer: c. Notify the site safety officer immediately.
Explanation: Immediate notification to the site safety officer is crucial, allowing for the prompt procurement of the required SDS and ensuring worker safety.

348. What information is crucial to include when labeling a hazardous chemical container?
a. Chemical Composition.
b. Manufacturer's Name and Address.
c. Signal word, Pictogram, Hazard Statement, and Precautionary Statement.
d. The Shelf Life of the Chemical.

Answer: c. Signal word, Pictogram, Hazard Statement, and Precautionary Statement.
Explanation: Including a signal word, pictograms, hazard and precautionary statements on labels is crucial to convey the risks and safety measures swiftly and effectively.

349. When a chemical hazard is non-routine, temporary, and occurs only in specific areas, what method of communication is typically deemed acceptable?
a. Placing a label on the chemical container.
b. Creating a new Safety Data Sheet (SDS).
c. Providing specific training for affected employees.
d. Sending a memo to all employees.

Answer: c. Providing specific training for affected employees.
Explanation: For non-routine, temporary chemical hazards, imparting specific training to employees who may be exposed is essential to ensure their understanding of the risks and precautions needed.

350. In the event of a construction site injury where the victim is unconscious but breathing, what should be the immediate action taken?
a. Commence CPR.
b. Place the victim in the recovery position.
c. Administer smelling salts to revive the victim.
d. Shake the victim to attempt to wake them up.

Answer: b. Place the victim in the recovery position.
Explanation: When a person is unconscious but breathing, placing them in the recovery position is crucial to keep their airway clear and open, preventing further complications until medical help arrives.

351. During an electrical incident, a worker has suffered a burn. What should be the first step in administering first aid?
a. Apply ice to the burn.
b. Cover the burn with a plastic wrap.
c. Apply butter to the burn.
d. Cool the burn with running water.

Answer: d. Cool the burn with running water.
Explanation: Cooling the burn under running water is the immediate first aid action to reduce pain and damage, after ensuring the victim is no longer in contact with the electrical source.

352. When a worker experiences a chemical splash in the eye, how long should the eye be flushed with water?
a. Until the pain subsides.
b. For at least 15 minutes.
c. For 5 minutes.
d. Until the eye is no longer red.

Answer: b. For at least 15 minutes.
Explanation: Flushing the eye continuously with water for at least 15 minutes is critical to dilute and remove the chemical, minimizing damage to the eye tissue.

353. In the event of a fall from height, a worker is complaining of severe back pain. What is the appropriate first aid response?
a. Move the worker to a comfortable position.
b. Apply heat to the affected area.
c. Immobilize the worker and await professional medical assistance.
d. Encourage the worker to stretch the back to alleviate pain.

Answer: c. Immobilize the worker and await professional medical assistance.
Explanation: When dealing with potential spinal injuries, it is imperative to immobilize the affected individual and await professional medical assistance to prevent further injury.

354. A worker has sustained a penetrating object injury. What is the recommended first aid approach?
a. Remove the object immediately.
b. Apply pressure around the object to control bleeding.
c. Push the object further to reduce bleeding.
d. Twist the object to seal the wound.

Answer: b. Apply pressure around the object to control bleeding.
Explanation: When an individual is impaled by an object, applying pressure around the object to control bleeding, without removing it, is crucial as removal can lead to more damage and bleeding.

355. During site excavation, a worker was exposed to poisonous gas and is experiencing difficulty breathing. What should be the initial response?
a. Administer oxygen and keep the worker still.
b. Move the worker to fresh air immediately and monitor vital signs.
c. Perform chest compressions.
d. Ask the worker to hold their breath until help arrives.

Answer: b. Move the worker to fresh air immediately and monitor vital signs.
Explanation: Quick removal from exposure and moving to fresh air is crucial in cases of inhalation of poisonous gas, and monitoring vital signs is important while awaiting professional medical help.

356. In case of a severed limb, what is the priority in providing first aid?
a. Apply a tourniquet above the injury site.
b. Retrieve the severed part and place it directly on ice.
c. Clean the wound with hydrogen peroxide.
d. Control bleeding with direct pressure and elevate the injury.

Answer: d. Control bleeding with direct pressure and elevate the injury.
Explanation: Immediate control of bleeding by applying direct pressure and elevating the injury is paramount in such severe injuries to prevent shock and further complications.

357. A worker experiences a sudden, severe headache, confusion, and weakness in the left arm. Recognizing these as symptoms of a possible stroke, what should be the immediate action?
a. Lay the worker down and elevate the legs.
b. Give the worker aspirin and water.
c. Call emergency services immediately.
d. Ask the worker to shake it off and continue working.

Answer: c. Call emergency services immediately.
Explanation: Recognizing the symptoms of a stroke and calling for immediate professional medical assistance is critical in minimizing the damage and improving the prognosis.

358. For a construction worker showing signs of heat exhaustion such as heavy sweating and weakness, what should be the first aid response?
a. Give the worker a sports drink and have them rest in a cool place.
b. Ask the worker to continue working but at a slower pace.
c. Immerse the worker in cold water.
d. Provide hot beverages to the worker to balance the body temperature.

Answer: a. Give the worker a sports drink and have them rest in a cool place.
Explanation: Providing electrolyte-rich fluids and allowing the worker to rest in a cool place helps in recovering from heat exhaustion by replenishing lost fluids and reducing body temperature.

359. A worker has sustained a minor cut on the arm. After cleaning the wound, what is the next step in providing first aid?
a. Apply a tourniquet above the wound.
b. Cover the wound with a sterile dressing or bandage.
c. Leave the wound uncovered to let it breathe.
d. Apply hydrogen peroxide to the wound daily.

Answer: b. Cover the wound with a sterile dressing or bandage.
Explanation: Covering the cleaned wound with a sterile dressing or bandage is important to protect it from infection and contaminants, promoting healing.

360. A contractor is using a reciprocating saw to cut through metal piping. Which of the following should be done to ensure safety while using the saw?
a. Use the saw with one hand while holding the metal pipe with the other.
b. Use a dull blade to prevent sparks.
c. Secure the material being cut, and use both hands to operate the saw.
d. Avoid using personal protective equipment to maintain visibility.

Answer: c. Secure the material being cut, and use both hands to operate the saw.
Explanation: Securely clamping the material and using both hands to operate the saw ensures stability, reduces the risk of accidents, and allows for more controlled cuts.

361. When operating a pneumatic nail gun, it is crucial to:
a. Disable the safety mechanism to increase speed.
b. Use the highest possible pressure for maximum efficiency.
c. Never bypass or remove safety devices, and use the tool at the manufacturer's recommended pressure.
d. Keep fingers close to the trigger when not in use to be ready for the next nail.

Answer: c. Never bypass or remove safety devices, and use the tool at the manufacturer's recommended pressure.
Explanation: Maintaining safety devices and operating tools at recommended pressures prevents malfunction and reduces the risk of injury.

362. How often should a contractor inspect electrical cords and plugs on the construction site?
a. Weekly.
b. Monthly.
c. Daily, before use.
d. Annually.

Answer: c. Daily, before use.
Explanation: Inspecting electrical cords and plugs daily before use ensures any damages or faults are identified promptly, preventing electrical accidents.

363. A contractor opts to use a ladder to access a high location on the construction site. What is the appropriate angle at which the ladder should be positioned?
a. 45 degrees.
b. 60 degrees.
c. 75 degrees.
d. 90 degrees.

Answer: b. 60 degrees.
Explanation: A ladder should be positioned so it's one foot away from the wall for every four feet of height, which roughly corresponds to a 60-degree angle, minimizing the risk of tipping.

364. A welder must always use a face shield while welding to protect against:
a. Noise.
b. Dust.
c. Ultraviolet and infrared light.
d. Falling objects.

Answer: c. Ultraviolet and infrared light.
Explanation: Face shields protect welders from the harmful ultraviolet and infrared light produced during welding, preventing eye damage.

365. Which of the following practices is crucial when operating a table saw?
a. Standing directly behind the saw blade.
b. Using hands to remove offcuts from near the blade.
c. Using push sticks or push blocks when cutting small or narrow pieces.
d. Removing the blade guard for better visibility.

Answer: c. Using push sticks or push blocks when cutting small or narrow pieces.
Explanation: Using push sticks or push blocks minimizes the risk of hand contact with the saw blade, especially when cutting small or narrow pieces.

366. Which of the following is the safest way to lift heavy construction equipment manually?
a. Lift with the back straight, using the legs.
b. Lift quickly to minimize time spent carrying the load.
c. Twist the torso while lifting to increase range of motion.
d. Bend at the waist and lift with the back.

Answer: a. Lift with the back straight, using the legs.
Explanation: Lifting with the back straight and using the legs to lift reduces strain on the back, reducing the risk of injury.

367. During demolition, which of the following is the safest method to avoid injuries from flying debris?
a. Use handheld shields.
b. Stand as close to the demolition point as possible.
c. Use protective screens or barricades.
d. Rely on personal protective equipment only.

Answer: c. Use protective screens or barricades. Explanation: Utilizing protective screens or barricades effectively contains flying debris, minimizing the risk of injury to personnel.

368. While operating a power drill, a worker notices a strange smell and smoke coming from the tool. What should be the immediate action?
a. Continue to use the drill but at a slower pace.
b. Discontinue use immediately and unplug the tool, if applicable.
c. Ignore the smell and smoke; it's likely due to overuse.
d. Use a fan to dissipate the smoke and continue working.

Answer: b. Discontinue use immediately and unplug the tool, if applicable.
Explanation: If a power tool shows signs of malfunction such as smoke or strange smells, it should be immediately turned off and unplugged to prevent further damage or potential injury.

369. A construction worker, while operating a handheld circular saw, needs to ensure that:
a. The saw blade is fully stopped before setting it down.
b. The guard is tied back for increased ease of use.
c. The saw is used with one hand to maintain balance.
d. The blade is removed after every cut to prevent accidents.

Answer: a. The saw blade is fully stopped before setting it down.
Explanation: Ensuring the saw blade is fully stopped before setting it down prevents unintended contact with moving parts, reducing the risk of cuts or lacerations.

370. In a construction site scenario where flammable materials are abundant, which type of fire extinguisher is most appropriate?
a. Water extinguisher
b. CO2 extinguisher
c. Dry chemical extinguisher
d. Wet chemical extinguisher

Answer: c. Dry chemical extinguisher
Explanation: Dry chemical extinguishers are versatile and suitable for most fire types, including those involving flammable liquids and gases, making them ideal for construction sites with an abundance of flammable materials.

371. A contractor is responsible for implementing fire safety measures at a construction site. Which of the following is a primary requirement?
a. The availability of open flame devices
b. Ensuring all exits are kept locked
c. Installation of smoke detectors and fire alarms
d. Storing flammable liquids near exit routes

Answer: c. Installation of smoke detectors and fire alarms
Explanation: Installing smoke detectors and fire alarms is crucial for early detection of fires, allowing for quick evacuation and response, minimizing damage and injury.

372. In high-rise construction projects, what is essential for fire protection during construction phases?
a. Ignition source control
b. Installation of temporary stairs
c. Immediate removal of fire extinguishers after completion
d. Use of flammable building materials

Answer: a. Ignition source control
Explanation: Controlling ignition sources is paramount in high-rise construction projects to prevent the outbreak of fire during construction phases when the building is most vulnerable.

373. For a construction project involving welding operations, which precaution is critical to avoid fires?
a. Perform operations in confined spaces
b. Store flammable materials close to welding operations
c. Ensure availability of fire watch personnel during and after operations
d. Use of water extinguishers only

Answer: c. Ensure availability of fire watch personnel during and after operations
Explanation: Fire watch personnel can immediately detect and respond to fires during and after welding operations, ensuring any fires are dealt with promptly and effectively.

374. When using a fuel-powered generator at a construction site, what is a crucial safety measure to prevent fire?
a. Refueling while the generator is running
b. Placing the generator near combustible materials
c. Allowing adequate ventilation and distance from combustibles
d. Overloading the generator for maximum efficiency

Answer: c. Allowing adequate ventilation and distance from combustibles
Explanation: Proper ventilation reduces the risk of accumulation of toxic fumes, and maintaining distance from combustibles prevents the ignition of materials due to heat.

375. To manage fire risk from hot work activities effectively, which step is crucial?
a. Obtaining a hot work permit before starting the operation
b. Performing hot work near flammable substances
c. Avoiding the use of fire blankets and shields
d. Ignoring fire watches for minor hot work activities

Answer: a. Obtaining a hot work permit before starting the operation
Explanation: A hot work permit ensures that all fire prevention measures are in place before beginning any activities that involve open flames or produce heat or sparks.

376. When storing flammable liquids on a construction site, which of the following is most critical?
a. Keeping them in open containers
b. Storing them near sources of ignition
c. Using approved containers and tanks
d. Mixing different types of flammable liquids in one container

Answer: c. Using approved containers and tanks
Explanation: Using approved containers and tanks for flammable liquids reduces the risk of leaks and spills, minimizing the chance of fire outbreaks.

377. A temporary construction building is installed on site. What fire safety measure is essential?
a. Blocking all exits with materials
b. Installing temporary heating equipment close to combustibles
c. Equipping the building with suitable fire extinguishers
d. Avoiding fire drills and evacuation plans

Answer: c. Equipping the building with suitable fire extinguishers
Explanation: Having suitable fire extinguishers ensures that small fires can be controlled, reducing potential damage and risk to life.

378. A construction project is located in a region prone to wildfires. Which action is pivotal for fire prevention?
a. Ignoring local fire danger warnings
b. Creating defensible space around the construction area
c. Storing construction waste in the open
d. Discarding cigarette butts around the construction area

Answer: b. Creating defensible space around the construction area
Explanation: Defensible space acts as a buffer zone, reducing the risk of wildfires reaching the construction site and providing a safer environment for firefighters to operate if needed.

379. During excavation work, a gas line is struck, resulting in a gas leak. Which immediate action is essential to prevent a fire?
a. Ignite the gas to burn it off
b. Continue working while avoiding the leak area
c. Evacuate the area and contact the appropriate emergency services
d. Cover the leak with soil to block the gas

Answer: c. Evacuate the area and contact the appropriate emergency services
Explanation: Evacuating the area ensures the safety of all personnel, and contacting emergency services allows for the prompt addressing of the gas leak, preventing potential fire outbreaks.

380. When working on an electrical installation in a damp environment, what equipment is crucial for a contractor to use?
a. Non-insulated tools
b. Ground-fault circuit interrupter (GFCI)
c. Conductive shoes
d. Metal ladder

Answer: b. Ground-fault circuit interrupter (GFCI)
Explanation: A GFCI is crucial in damp environments to prevent electric shock hazards by interrupting the electrical circuit when it detects an imbalance in the electrical current.

381. A contractor is conducting a renovation which involves exposure to live wires. What precautionary measure is essential?
a. Using metallic tape for insulation
b. Using damaged insulated tools
c. Implementing lockout/tagout procedures
d. Handling live wires with wet hands

Answer: c. Implementing lockout/tagout procedures
Explanation: Lockout/tagout procedures ensure that the electrical energy is isolated, preventing unintentional energization and protecting workers from electrical shock.

382. During a project involving high-voltage systems, which personal protective equipment (PPE) is indispensable?
a. Non-flame-resistant clothing
b. Insulated gloves and sleeves
c. Metal head protection
d. Non-insulated boots

Answer: b. Insulated gloves and sleeves
Explanation: When working with high-voltage systems, using insulated gloves and sleeves is essential to protect against electrical shocks and burns.

383. For a construction project, which practice is crucial when using extension cords?
a. Using cords with damaged insulation
b. Running cords through water
c. Using cords rated for the load they are carrying
d. Daisy chaining multiple extension cords

Answer: c. Using cords rated for the load they are carrying
Explanation: Using extension cords that are appropriately rated prevents overheating and electrical fires, ensuring the safety of the construction site.

384. A contractor is installing electrical systems. Which step is pivotal to mitigate electrical hazards?
a. Ignoring manufacturer's instructions
b. Using mismatched breakers and panels
c. Conducting a detailed risk assessment
d. Leaving exposed wires unattended

Answer: c. Conducting a detailed risk assessment
Explanation: A detailed risk assessment identifies potential electrical hazards, allowing for the implementation of control measures to mitigate risks.

385. What is crucial when operating power tools to ensure electrical safety?
a. Using tools with damaged cords
b. Using tools in wet or damp conditions
c. Disconnecting the power source when changing accessories
d. By-passing the ground plug on three-prong plugs

Answer: c. Disconnecting the power source when changing accessories
Explanation: Disconnecting the power source is crucial to prevent accidental activation of the tool, which can lead to serious injuries.

386. Which method is crucial for ensuring the safety of temporary electrical installations on a construction site?
a. Using ungrounded systems
b. Regular inspection and maintenance
c. Overloading circuits with multiple devices
d. Using damaged or frayed cables

Answer: b. Regular inspection and maintenance
Explanation: Regular inspection and maintenance of temporary electrical installations are crucial to identify and rectify any deficiencies that could pose a risk.

387. When working near overhead power lines, maintaining a minimum safe distance is critical. What is the minimum distance that must be maintained from a 50 kV power line?
a. 5 feet
b. 10 feet
c. 15 feet
d. 20 feet

Answer: c. 15 feet
Explanation: OSHA specifies that for voltages over 50 kV, the minimum distance is 15 feet to prevent the risk of arcing and electrical shock.

388. Which safety practice is pivotal when a contractor is working on an energized electrical panel?
a. Not using any personal protective equipment
b. Standing on a wet surface
c. Using insulated tools
d. Removing panel covers without turning off the power

Answer: c. Using insulated tools
Explanation: Insulated tools are essential to prevent the risk of electrical shock when working on energized electrical panels.

389. In a scenario where a worker receives an electrical shock, what is the first course of action that should be taken by a bystander?
a. Attempt to pull the victim away with bare hands
b. Immediately touch the victim to check responsiveness
c. Disconnect the power source or use a non-conductive object to remove the victim from contact
d. Leave the victim and do not attempt to help until the arrival of medical personnel

Answer: c. Disconnect the power source or use a non-conductive object to remove the victim from contact. Explanation: Disconnecting the power source or using a non-conductive object is crucial to safely remove the victim from the electrical source, preventing further harm.

390. Before initiating a trench excavation, what is crucial for a contractor to identify and locate?
a. Site topography only
b. Underground utilities only
c. Nearby vegetation only
d. Both site topography and underground utilities

Answer: d. Both site topography and underground utilities
Explanation: Identifying both site topography and underground utilities is crucial to avoid damaging underground structures and to plan the excavation safely, reducing risks of cave-ins or collapses.

391. A contractor is excavating a trench deeper than 5 feet. What protective system is essential to implement to prevent a trench collapse?
a. Sloping only
b. Bench system only
c. Shoring system only
d. Either a sloping, benching, or shoring system based on soil conditions

Answer: d. Either a sloping, benching, or shoring system based on soil conditions. Explanation: Depending on the soil conditions, either sloping, benching, or a shoring system is necessary to prevent a trench collapse in trenches deeper than 5 feet.

392. In a scenario where a trench is 20 feet deep, how often should a competent person inspect the trench?
a. Once a week
b. At the start of each shift
c. Only after a rainstorm
d. Only after the installation of a support system

Answer: b. At the start of each shift
Explanation: A competent person should inspect the trench at the start of each shift and after any hazard-increasing event such as a rainstorm to ensure the continued safety of the excavation site.

393. When excavating near public roads, what safety measure is pivotal for contractors to implement?
a. Use of signal persons only
b. Use of barricades and warning signs only
c. Use of designated walkways only
d. Use of signal persons, barricades, warning signs, and designated walkways

Answer: d. Use of signal persons, barricades, warning signs, and designated walkways
Explanation: To ensure the safety of both workers and the public, implementing a combination of signal persons, barricades, warning signs, and designated walkways is essential.

394. For trenches 4 feet or deeper, what crucial provision must a contractor make?
a. Only ladder access
b. Only ramp access
c. Safe access and egress at intervals of 50 feet or less
d. Safe access and egress at intervals of 100 feet or less

Answer: c. Safe access and egress at intervals of 50 feet or less
Explanation: For trenches 4 feet or deeper, providing safe access and egress, such as ladders or ramps, at intervals of 50 feet or less is a crucial safety requirement.

395. What is the minimum distance a spoil pile must be kept from the edge of a trench?
a. 2 feet
b. 3 feet
c. 5 feet
d. 6 feet

Answer: a. 2 feet
Explanation: Spoil piles must be kept at least 2 feet from the edge of the trench to prevent materials from falling back into the trench and to avoid additional pressure on trench walls.

396. A contractor is using a hydraulic support system for trench excavation. What is vital for ensuring the safety of the workers inside the trench?
a. Leaving the hydraulic system disengaged
b. Removing the system before workers enter
c. Engaging the system while workers are inside
d. Installing the system only after the trench has reached its final depth

Answer: c. Engaging the system while workers are inside
Explanation: Engaging the hydraulic support system while workers are inside the trench is vital to provide adequate support to trench walls and prevent collapses.

397. In a scenario involving excavations near existing structures, what is critical for maintaining the stability of nearby buildings?
a. Using any protective system
b. Ignoring support systems for the structures
c. Providing support systems to avoid undermining the structures
d. Excavating closer to the structures for easier access

Answer: c. Providing support systems to avoid undermining the structures
Explanation: Providing support systems to the adjacent structures is critical to avoid undermining and compromising the stability of the structures.

398. When selecting protective systems for trenching and excavation, which factor is crucial to consider?
a. The color of the soil only
b. The cost of the protective system only
c. The type of soil and depth of the trench
d. The preferences of the workers only

Answer: c. The type of soil and depth of the trench
Explanation: The type of soil and the depth of the trench are crucial factors in selecting appropriate protective systems to ensure the stability of the excavation.

399. For an excavation project in a densely populated area, what is paramount to ensure the safety of the general public?
a. Removing safety barriers during work hours
b. Installing adequate barriers, signs, and coverings
c. Allowing public access to the excavation site
d. Not informing local authorities about the excavation project

Answer: b. Installing adequate barriers, signs, and coverings
Explanation: Installing adequate barriers, signs, and coverings is paramount to prevent unauthorized access and to protect the general public from potential hazards associated with the excavation project.

400. When incorporating innovative green building materials in a project, what is paramount for contractors to consider?
a. Aesthetic appeal only
b. Cost-effectiveness only
c. Structural integrity and compliance with building codes
d. Popularity among clients only

Answer: c. Structural integrity and compliance with building codes
Explanation: While innovative and sustainable, green building materials must also comply with building codes and maintain structural integrity to ensure the safety and longevity of the construction project.

401. In constructing a building with passive solar design, which aspect is crucial for a contractor to optimize?
a. Window size only
b. Shading and overhangs only
c. Orientation and window placement
d. Color of the exterior paint only

Answer: c. Orientation and window placement
Explanation: Optimizing the orientation and window placement is essential in passive solar design to maximize solar heat gain in winter and minimize it in summer, achieving energy efficiency.

402. In a scenario involving the restoration of a historic building, what is essential for a contractor to maintain?
a. Only the original color scheme
b. Only the original façade
c. The original architectural integrity and historical value
d. Only the original interior layout

Answer: c. The original architectural integrity and historical value
Explanation: Maintaining the original architectural integrity and historical value is vital in historic building restoration to preserve cultural heritage and comply with preservation standards.

403. In a design-build project delivery method, which entity is the contractor contractually obligated to?
a. The designer only
b. The owner only
c. The subcontractor only
d. The project manager only

Answer: b. The owner only
Explanation: In design-build, the contractor is contractually obligated to the owner, providing a single point of responsibility for both design and construction phases of the project.

404. For a contractor specializing in modular construction, what is crucial in ensuring project success?
a. Ignoring building codes
b. Prioritizing on-site construction
c. Precise planning and coordination with manufacturers
d. Utilizing traditional construction methods only

Answer: c. Precise planning and coordination with manufacturers
Explanation: Precise planning and coordination with manufacturers are critical in modular construction to ensure that prefabricated components meet specifications and are delivered on time.

154

405. In a construction project involving structural steel, what is a crucial factor a contractor must consider during erection?
a. Ignoring the sequential erection plan
b. Not utilizing temporary bracing
c. Strict adherence to the erection sequence and bracing requirements
d. Not consulting the structural engineer

Answer: c. Strict adherence to the erection sequence and bracing requirements
Explanation: Adhering strictly to the erection sequence and bracing requirements is vital to maintain structural stability during the erection of structural steel.

406. When implementing Value Engineering in a project, what is the main objective for a contractor?
a. To increase the project cost
b. To optimize the project's cost-effectiveness without compromising functionality and quality
c. To reduce the quality of materials
d. To prolong the project timeline

Answer: b. To optimize the project's cost-effectiveness without compromising functionality and quality
Explanation: The main objective of Value Engineering is to optimize cost-effectiveness, enhance functionality, and maintain quality, improving overall project value.

407. In managing a Lean Construction project, what is pivotal for contractors to minimize?
a. Communication with stakeholders
b. Construction waste and inefficiencies
c. Usage of innovative construction methods
d. Collaboration among project teams

Answer: b. Construction waste and inefficiencies
Explanation: In Lean Construction, minimizing construction waste and inefficiencies is paramount to optimize resource utilization and improve project delivery.

408. When constructing in a seismic zone, what construction practice is crucial for contractors to integrate?
a. Rigidity in structural components
b. Ductility in structural components
c. Cost minimization only
d. Aesthetic enhancements only

Answer: b. Ductility in structural components
Explanation: Incorporating ductility in structural components is crucial in seismic zones to allow buildings to deform and absorb energy without collapsing during seismic activity.

409. In constructing high-performance buildings, what aspect is crucial for contractors to focus on?
a. Aesthetic appeal only
b. Energy efficiency and sustainability
c. Using the cheapest available materials
d. Minimizing the number of windows

Answer: b. Energy efficiency and sustainability
Explanation: Focusing on energy efficiency and sustainability is crucial when constructing high-performance buildings to minimize environmental impact and reduce energy consumption.

410. In a high-rise construction project, what is crucial to maintain when selecting the curtain wall system?
a. Aesthetic Preference Only
b. Structural Integrity and Weather Tightness
c. Color Consistency Only
d. Manufacturer Reputation Only

Answer: b. Structural Integrity and Weather Tightness
Explanation: When selecting curtain wall systems, maintaining structural integrity and ensuring weather tightness is crucial for the longevity and safety of high-rise buildings, preventing water infiltration and structural damages.

411. When employing top-down construction methods in high-rise buildings, what is a contractor primarily seeking to achieve?
a. Prolonging the construction schedule
b. Reduction in construction timeline
c. Maximization of labor costs
d. Reduction in construction quality

Answer: b. Reduction in construction timeline
Explanation: Top-down construction methods are primarily utilized to reduce the overall construction timeline by allowing simultaneous construction of the superstructure and the substructure.

412. In the construction of a high-rise building, what is the primary purpose of using a tuned mass damper?
a. To decrease the building's weight
b. To increase the building's aesthetic appeal
c. To mitigate building sway due to wind and seismic activity
d. To increase the building's height

Answer: c. To mitigate building sway due to wind and seismic activity
Explanation: Tuned mass dampers are used to mitigate the sway of high-rise buildings caused by wind loads or seismic activity, enhancing the building's stability and comfort for occupants.

413. When utilizing slipform construction technique in high-rise buildings, what is a key advantage a contractor aims to achieve?
a. Reduction in construction speed
b. Continuous, uninterrupted concrete pour
c. Increase in the use of formwork
d. Maximization of construction errors

Answer: b. Continuous, uninterrupted concrete pour
Explanation: The slipform construction technique allows for a continuous, uninterrupted concrete pour, which is efficient for constructing vertical elements like cores and shear walls in high-rise buildings.

414. In a scenario involving the construction of a high-rise in a dense urban environment, what is crucial for a contractor to manage effectively?
a. Ignoring logistical constraints
b. Overlooking local building codes
c. Logistical planning and coordination with local authorities
d. Ignoring noise and disruption concerns

Answer: c. Logistical planning and coordination with local authorities. Explanation: Managing logistics and coordinating with local authorities is crucial in dense urban environments to mitigate traffic, noise, and disruption and to comply with local ordinances and building codes.

415. In the construction of high-rise buildings, what is a crucial factor when considering the use of structural steel frames?
a. Ignoring fire protection measures
b. Fire protection and the use of fire-resistant materials
c. Minimizing the number of connections
d. Utilizing untreated steel only

Answer: b. Fire protection and the use of fire-resistant materials
Explanation: When utilizing structural steel frames in high-rises, incorporating fire protection measures and using fire-resistant materials are critical to prevent structural failure in the event of a fire.

416. During the construction of a high-rise, which practice is essential in ensuring efficient and safe vertical transportation of personnel and materials?
a. Ignoring the placement of construction elevators
b. Placement and maintenance of construction elevators
c. Minimizing the use of construction elevators
d. Using a single construction elevator

Answer: b. Placement and maintenance of construction elevators
Explanation: Proper placement and regular maintenance of construction elevators are essential to ensure the safe and efficient vertical transportation of personnel and materials in high-rise construction.

417. For a high-rise building with a deep foundation, what is the contractor primarily concerned with during the piling operation?
a. Ignoring soil conditions
b. Soil conditions and load-bearing capacity
c. Aesthetic of the piles
d. The color of the piles

Answer: b. Soil conditions and load-bearing capacity
Explanation: During piling operations for deep foundations in high-rises, contractors are primarily concerned with soil conditions and ensuring that piles have adequate load-bearing capacity to support the structure.

418. When constructing a mixed-use high-rise building, what is crucial for a contractor to integrate?
a. Disjointed building systems
b. A cohesive and integrated building system design
c. Ignoring mechanical, electrical, and plumbing systems integration
d. Isolated design considerations for different uses

Answer: b. A cohesive and integrated building system design
Explanation: A cohesive and integrated building system design is crucial in mixed-use high-rise buildings to ensure the seamless functioning of different uses and the efficiency of building systems.

419. In a high-rise construction project utilizing BIM, what is a key advantage a contractor aims to gain?
a. Increase in construction inaccuracies
b. Enhanced coordination and clash detection
c. Ignoring project stakeholders
d. Decreased project visualization

Answer: b. Enhanced coordination and clash detection
Explanation: Utilizing BIM in high-rise construction enables enhanced coordination among project stakeholders and effective clash detection, minimizing errors and rework during construction.

420. In historical restoration, when working with lime mortars, what is a critical consideration to ensure the longevity and integrity of the restoration?
a. Ignoring the curing time
b. Selecting a mortar with a higher compressive strength than the original
c. Adequate curing and compatible compressive strength with the original structure
d. Selecting a mortar with different composition from the original

Answer: c. Adequate curing and compatible compressive strength with the original structure
Explanation: When working with lime mortars, it is critical to allow adequate curing time and to choose a mortar with compatible compressive strength with the original structure to prevent damage and ensure longevity.

421. While restoring a historical building with deteriorated wooden elements, what is a pivotal step to maintain authenticity?
a. Replacing all wood with modern materials
b. Matching the wood species and using traditional joinery methods
c. Using a different wood species and modern joinery methods
d. Ignoring the original wood species and joinery methods used

Answer: b. Matching the wood species and using traditional joinery methods
Explanation: Matching the original wood species and using traditional joinery methods are pivotal to maintaining the historical accuracy and authenticity of the restored building.

422. When restoring a heritage building with stained glass windows, what is crucial to maintaining the integrity of the original design?
a. Completely replacing the stained glass with new materials
b. Ignoring the original colors and patterns
c. Careful documentation and replication of original colors and patterns
d. Using a modern design that contrasts the original

Answer: c. Careful documentation and replication of original colors and patterns
Explanation: When restoring stained glass, maintaining the integrity of the original design involves careful documentation and replication of the original colors, patterns, and lead came techniques used.

423. In the restoration of a historic building façade, what is the primary consideration for selecting a cleaning method?
a. Choosing the most aggressive method available
b. Ignoring the type of dirt or soiling present
c. Considering the type of soiling and substrate material
d. Prioritizing cost over substrate preservation

Answer: c. Considering the type of soiling and substrate material
Explanation: Selecting an appropriate cleaning method requires considering the type of soiling present and the substrate material to avoid damaging the historic façade.

424. While undertaking a historic restoration project involving masonry structures, what is essential to prevent further deterioration?
a. Using the hardest mortar available
b. Ignoring the porosity of the replacement materials
c. Matching the porosity and compressive strength of the original materials
d. Selecting materials based solely on color match

Answer: c. Matching the porosity and compressive strength of the original materials
Explanation: To prevent further deterioration, it's essential to match the porosity and compressive strength of the original materials when restoring masonry structures.

425. In a scenario where a contractor is restoring a historic site with mural paintings, which action is critical to maintaining the artistic value?
a. Overpainting the entire mural
b. Ignoring the original color palette used
c. Conducting a thorough analysis and utilizing conservation techniques
d. Changing the theme of the mural

Answer: c. Conducting a thorough analysis and utilizing conservation techniques
Explanation: Maintaining the artistic value of mural paintings involves conducting a thorough analysis of the original materials and utilizing careful conservation techniques to preserve the existing artwork.

426. When restoring a historic building, why is it crucial to perform a detailed historical research and documentation of the original structure?
a. To change the original design elements
b. To ignore the original construction methods used
c. To ensure accurate restoration to the original state
d. To introduce contrasting modern elements

Answer: c. To ensure accurate restoration to the original state
Explanation: Detailed historical research and documentation of the original structure are crucial to ensure the accurate restoration and replication of original design elements, materials, and construction methods.

427. While restoring a historic building, what is the key consideration when dealing with deteriorated structural elements?
a. Replacing with different materials without analysis
b. Ignoring the structural integrity of the remaining elements
c. Comprehensive assessment and sympathetic integration of new elements
d. Reducing the load-bearing capacity of the structure

Answer: c. Comprehensive assessment and sympathetic integration of new elements
Explanation: A comprehensive assessment of the remaining elements and sympathetic integration of new elements are key to maintaining structural integrity while respecting the historic fabric of the building.

428. In a historic restoration project involving intricate decorative plasterwork, what is a vital step to preserve the original craftsmanship?
a. Creating molds of existing work for replication
b. Ignoring the detailing and simplifying the design
c. Replacing intricate work with plain finishes
d. Using contrasting modern decorative elements

Answer: a. Creating molds of existing work for replication
Explanation: To preserve the original craftsmanship in decorative plasterwork, creating molds of existing work for accurate replication is a vital step in maintaining the historical integrity of the design.

429. In restoring a historical building with lead-based paint, what is crucial for the contractor to consider for the safety of occupants and workers?
a. Ignoring safety protocols and regulations
b. Employing unsafe removal methods
c. Implementing lead-safe work practices and compliance with regulations
d. Leaving deteriorated lead paint untreated

Answer: c. Implementing lead-safe work practices and compliance with regulations
Explanation: When dealing with lead-based paint, implementing lead-safe work practices and compliance with regulations is crucial to mitigate the risk of lead exposure to occupants and workers.

430. In constructing a building using modular construction, what is crucial for ensuring the structural integrity of the assembled modules?
a. Ignoring the module-to-module connections
b. Reducing the number of fasteners at connection points
c. Precise fabrication and robust inter-module connections
d. Utilizing inconsistent fabrication techniques for each module

Answer: c. Precise fabrication and robust inter-module connections
Explanation: Ensuring structural integrity in modular construction requires precise fabrication and robust inter-module connections to effectively transfer loads and resist lateral forces.

431. When selecting materials for a prefab construction project, what is a critical consideration to optimize transportation logistics?
a. Selecting the heaviest materials available
b. Ignoring the dimensions and weight of the prefab components
c. Prioritizing lightweight and compact materials
d. Using oversized components without considering transport limitations

Answer: c. Prioritizing lightweight and compact materials
Explanation: For prefab construction, prioritizing lightweight and compact materials is crucial to optimizing transportation logistics, minimizing costs, and reducing the risk of transport-related damages.

432. In a scenario where a contractor is assembling a modular building, what is a primary concern to ensure the longevity and performance of the building envelope?
a. Ignoring the sealing of joints between modules
b. Utilizing non-durable sealing materials
c. Ensuring airtight and watertight seals at module junctions
d. Leaving gaps between modules to allow for ventilation

Answer: c. Ensuring airtight and watertight seals at module junctions
Explanation: Ensuring airtight and watertight seals at module junctions is vital to prevent water ingress and air leakage, maintaining the building's thermal performance and durability.

433. When designing a prefab building, what is a pivotal aspect to facilitate ease of assembly and disassembly?
a. Using irreversible connection methods
b. Ignoring the adaptability and flexibility of the design
c. Incorporating standardized components and reversible connections
d. Designing each component with unique, non-standard dimensions

Answer: c. Incorporating standardized components and reversible connections
Explanation: Incorporating standardized components and reversible connections in the design of prefab buildings facilitates ease of assembly and disassembly, allowing for adaptability and reusability of components.

434. What is a crucial consideration when planning the site for a modular construction project to ensure smooth installation of modules?
a. Ignoring the access routes and site constraints
b. Choosing a site with limited space and accessibility
c. Assessing site accessibility, space, and ground conditions
d. Selecting a site with poor ground conditions

Answer: c. Assessing site accessibility, space, and ground conditions
Explanation: A thorough assessment of site accessibility, available space, and ground conditions is crucial in planning the site for a modular construction project to ensure smooth installation and avoid delays.

435. In the case of a modular construction project in a seismic zone, what is a fundamental design consideration to enhance seismic resilience?
a. Reducing the lateral load resisting elements
b. Ignoring the seismic forces acting on the structure
c. Incorporating ductile detailing and adequate lateral load resisting systems
d. Designing rigid, non-flexible structural connections

Answer: c. Incorporating ductile detailing and adequate lateral load resisting systems
Explanation: For modular construction in seismic zones, incorporating ductile detailing and adequate lateral load resisting systems are fundamental to enhancing seismic resilience and ensuring the safety of the structure.

436. When implementing prefab construction for a residential project, what is essential to ensure efficient integration of mechanical, electrical, and plumbing (MEP) systems?
a. Ignoring coordination between different trades
b. Installing MEP systems without considering spatial constraints
c. Detailed coordination and integration of MEP systems during the design phase
d. Utilizing oversized MEP components without considering space availability

Answer: c. Detailed coordination and integration of MEP systems during the design phase
Explanation: Detailed coordination and integration of MEP systems during the design phase are essential in prefab construction to ensure efficient utilization of space and avoid conflicts during assembly.

437. In a modular construction project, what is a vital step to maintain quality control and ensure compliance with building standards?
a. Bypassing inspections and testing
b. Ignoring manufacturing tolerances and quality standards
c. Conducting rigorous inspections and testing during the fabrication phase
d. Using substandard materials and components

Answer: c. Conducting rigorous inspections and testing during the fabrication phase
Explanation: Maintaining quality control in modular construction involves conducting rigorous inspections and testing during the fabrication phase to ensure compliance with building standards and identify any defects or non-conformities.

438. When managing a prefab construction project, what is crucial for optimizing the construction schedule and avoiding delays?
a. Ignoring the lead times for prefab components
b. Overlapping the design, fabrication, and site preparation phases
c. Waiting for the completion of design before starting fabrication
d. Delaying site preparation until the arrival of prefab components

Answer: b. Overlapping the design, fabrication, and site preparation phases
Explanation: Optimizing the construction schedule in prefab construction involves overlapping the design, fabrication, and site preparation phases, allowing for parallel progression of activities and reducing the overall project duration.

439. For a contractor involved in a prefab construction project, what is critical for ensuring the project's success and client satisfaction?
a. Ignoring client requirements and preferences
b. Compromising on quality to reduce costs
c. Clear communication and meticulous planning from the project's inception
d. Implementing changes without client approval

Answer: c. Clear communication and meticulous planning from the project's inception
Explanation: Clear communication with all stakeholders and meticulous planning from the project's inception are critical in prefab construction to align expectations, manage complexities, and ensure the successful delivery of the project.

440. When retrofitting a historic masonry building located in a high seismic zone, which technique is pivotal to enhance the lateral load resistance of the structure?
a. Removing existing lateral load resisting elements
b. Ignoring the bracing of unreinforced masonry walls
c. Implementing base isolation systems
d. Avoiding the addition of shear walls

Answer: c. Implementing base isolation systems
Explanation: Base isolation systems are crucial in retrofitting historic masonry buildings in high seismic zones as they decouple the structure from ground motion, significantly reducing seismic forces and enhancing lateral load resistance.

441. In a project involving seismic retrofitting of a hospital, what design approach is crucial to maintain the building's functionality post-earthquake?
a. Designing for minimum seismic code requirements
b. Ignoring non-structural components
c. Employing performance-based seismic design
d. Avoiding the consideration of dynamic analysis

Answer: c. Employing performance-based seismic design
Explanation: Employing performance-based seismic design is crucial for maintaining the functionality of critical buildings like hospitals post-earthquake by addressing both structural and non-structural components and ensuring that they meet specific performance objectives.

442. When conducting a seismic retrofit of a multi-story residential building, what is a critical consideration to improve the ductility of the structure?
a. Incorporating rigid diaphragm
b. Reducing the number of shear walls
c. Implementing steel moment-resisting frames
d. Avoiding flexible diaphragm

Answer: c. Implementing steel moment-resisting frames
Explanation: Steel moment-resisting frames are critical for improving the ductility of a multi-story building during seismic retrofitting as they allow for controlled displacement and deformation under seismic loads, reducing the likelihood of catastrophic failure.

443. In a scenario where a bridge requires seismic retrofitting, what technique is paramount for enhancing the bridge's seismic performance?
a. Ignoring soil-structure interaction
b. Removing expansion joints
c. Strengthening the bridge piers and abutments
d. Avoiding seismic isolation bearings

Answer: c. Strengthening the bridge piers and abutments
Explanation: Strengthening bridge piers and abutments is paramount in seismic retrofitting of bridges as it enhances the seismic performance and resilience of the bridge by improving its ability to withstand and dissipate seismic forces.

444. When retrofitting a building with soft story condition, what is an effective technique to mitigate the risk of collapse during an earthquake?
a. Reducing the stiffness of the soft story
b. Implementing shear walls or braced frames at the soft story level
c. Ignoring the reinforcement of the soft story
d. Avoiding the addition of infill walls

Answer: b. Implementing shear walls or braced frames at the soft story level
Explanation: Implementing shear walls or braced frames at the soft story level is an effective technique to mitigate the risk of collapse by providing additional lateral load resistance and reducing deformation and displacement in that story.

445. In a seismic retrofit project involving an unreinforced masonry building, which of the following is critical for improving the out-of-plane stability of masonry walls?
a. Removing wall anchors
b. Installing vertical and horizontal wall anchors
c. Reducing wall thickness
d. Avoiding the use of fiber-reinforced polymers

Answer: b. Installing vertical and horizontal wall anchors
Explanation: Installing vertical and horizontal wall anchors is critical in seismic retrofit projects involving unreinforced masonry buildings for improving the out-of-plane stability of masonry walls by effectively transferring seismic loads to the structure's lateral load resisting system.

446. In a project involving the seismic retrofitting of a concrete frame building, what method is crucial to enhance the lateral load capacity and ductility of the frame?
a. Implementing external post-tensioning
b. Ignoring the reinforcement of beam-column joints
c. Reducing the sectional dimensions of the columns
d. Avoiding the use of steel jacketing

Answer: a. Implementing external post-tensioning
Explanation: Implementing external post-tensioning in concrete frame buildings is crucial during seismic retrofitting as it enhances lateral load capacity and ductility by applying a pre-compressive force to the frame, reducing the likelihood of brittle failure.

447. When addressing seismic vulnerabilities of a school building, what is essential to protect non-structural elements and building contents during an earthquake?
a. Ignoring the anchorage of non-structural elements
b. Implementing seismic bracing and anchorage of non-structural elements
c. Avoiding the securing of building contents
d. Reducing the weight of building contents

Answer: b. Implementing seismic bracing and anchorage of non-structural elements
Explanation: Implementing seismic bracing and anchorage of non-structural elements is essential to protect them and building contents during an earthquake, preventing injuries and damage and maintaining the building's post-earthquake functionality.

448. In a seismic retrofit of a building with a large open space on the ground floor, what technique is essential to avoid a soft story collapse?
a. Implementing a rigid diaphragm on the ground floor
b. Avoiding the installation of additional shear walls or columns
c. Strengthening the ground floor with additional shear walls or columns
d. Reducing the stiffness of upper floors

Answer: c. Strengthening the ground floor with additional shear walls or columns
Explanation: Strengthening the ground floor with additional shear walls or columns is essential in seismic retrofit projects to avoid a soft story collapse by providing the required lateral stiffness and strength to resist seismic forces.

449. In seismic retrofitting of an industrial facility, which approach is critical to maintain the operational continuity of the facility post-earthquake?
a. Designing for only structural components
b. Ignoring the strengthening of equipment and machinery supports
c. Employing a holistic retrofit approach considering both structural and non-structural components
d. Avoiding the anchorage of equipment and machinery

Answer: c. Employing a holistic retrofit approach considering both structural and non-structural components
Explanation: Employing a holistic retrofit approach that considers both structural and non-structural components is critical to maintain the operational continuity of an industrial facility post-earthquake by addressing the vulnerabilities of equipment, machinery, and their supports, in addition to the structural components.

450. While retrofitting an old theater, the contractor is faced with addressing reverberation issues to improve speech intelligibility. Which acoustical solution is critical to implement?
a. Reducing the amount of absorbent materials
b. Ignoring diffusion strategies
c. Incorporating sound-reflective materials
d. Installing diffusers and absorbent materials

Answer: d. Installing diffusers and absorbent materials
Explanation: To address reverberation issues and improve speech intelligibility in a theater, it is essential to install diffusers and absorbent materials. These solutions help in controlling reflected sound, reducing reverberation time, and improving the overall acoustic quality of the space.

451. In a high-rise residential project, what is a pivotal acoustical solution to mitigate flanking transmission between adjacent units?
a. Incorporating single stud walls with direct attachment of gypsum board
b. Avoiding insulation in the partition walls
c. Employing resilient channel installations and double stud walls
d. Reducing the mass of the partition walls

Answer: c. Employing resilient channel installations and double stud walls
Explanation: To mitigate flanking transmission, employing resilient channel installations and constructing double stud walls are critical, as they break the direct path of sound transmission and significantly improve the Sound Transmission Class (STC) rating of the partition.

452. When constructing a recording studio, which of the following is crucial to minimize structure-borne sound transmission?
a. Ignoring isolation of mechanical equipment
b. Implementing a room-within-a-room construction
c. Reducing mass of the structure
d. Avoiding the use of damping materials

Answer: b. Implementing a room-within-a-room construction
Explanation: In a recording studio, implementing a room-within-a-room construction is crucial to minimize structure-borne sound transmission. This technique decouples the inner structure from the outer structure, reducing the transmission of vibrational energy into the structure.

453. In a hospital construction project, what is essential to maintain speech privacy in patient rooms?
a. Installing doors with high gaps and voids
b. Employing sound masking systems
c. Avoiding the use of acoustic seals on doors
d. Reducing the surface mass of partition walls

Answer: b. Employing sound masking systems
Explanation: In hospitals, employing sound masking systems is essential to maintain speech privacy in patient rooms. Sound masking introduces a background sound, reducing the intelligibility of human speech and helping in maintaining the confidentiality of conversations.

454. For a high-traffic airport terminal, which acoustical solution is paramount to control noise levels and reduce stress among travelers?
a. Reducing the number of sound-absorbing materials
b. Avoiding the installation of acoustic ceilings
c. Installing high NRC (Noise Reduction Coefficient) materials
d. Ignoring the layout of the space

Answer: c. Installing high NRC (Noise Reduction Coefficient) materials
Explanation: Installing materials with a high Noise Reduction Coefficient (NRC) is paramount in high-traffic areas like airport terminals to control noise levels. These materials absorb sound effectively, reducing the overall noise within the space and subsequently reducing stress among travelers.

455. In a project involving the construction of a library, what acoustical solution is crucial to maintain a quiet environment for reading and studying?
a. Avoiding the use of carpeting and draperies
b. Implementing HVAC systems without silencers
c. Incorporating acoustic panels and soft furnishings
d. Reducing the number of bookshelves to open up the space

Answer: c. Incorporating acoustic panels and soft furnishings
Explanation: For a library construction project, incorporating acoustic panels and soft furnishings is crucial. These solutions absorb sound energy, reducing noise levels and reverberation, and helping in maintaining a conducive environment for reading and studying.

456. In the construction of an open office layout, which acoustical solution is crucial to reduce the distraction and enhance concentration among employees?
a. Avoiding the use of acoustic furniture
b. Implementing sound-reflecting partitions
c. Reducing the height of office partitions
d. Employing sound-absorbing office furniture and partitions

Answer: d. Employing sound-absorbing office furniture and partitions
Explanation: Employing sound-absorbing office furniture and partitions is crucial in open office layouts. They help in reducing reflected sound and controlling noise levels, reducing distractions and enhancing concentration and productivity among employees.

457. When constructing an auditorium, what is a crucial consideration to ensure uniform sound distribution throughout the space?
a. Ignoring the shape of the ceiling
b. Incorporating splayed walls and a well-designed ceiling
c. Reducing the use of diffusers
d. Avoiding the positioning of loudspeakers

Answer: b. Incorporating splayed walls and a well-designed ceiling
Explanation: In auditorium construction, incorporating splayed walls and a well-designed ceiling is crucial. These design considerations help in distributing sound uniformly throughout the space, avoiding hotspots and dead zones, and ensuring a good listening experience for the audience.

458. In a museum construction project, what acoustical solution is essential to control the noise levels in exhibition halls with high footfall?
a. Reducing the absorption coefficient of materials
b. Installing hard and reflective surfaces
c. Employing high-impact noise barriers
d. Avoiding strategic placement of exhibits

Answer: c. Employing high-impact noise barriers
Explanation: In high-footfall exhibition halls, employing high-impact noise barriers is essential to control noise levels. These barriers block the transmission of sound between different areas, reducing the overall noise within the space and maintaining a pleasant environment for visitors.

459. During the construction of a multi-story parking garage adjacent to residential buildings, what acoustical solution is essential to mitigate the impact of noise on nearby residences?
a. Employing lightweight materials with less mass
b. Ignoring the use of acoustic screens
c. Incorporating noise barriers and acoustic screens
d. Avoiding the use of sound-absorbing materials on walls

Answer: c. Incorporating noise barriers and acoustic screens
Explanation: Incorporating noise barriers and acoustic screens is essential during the construction of a multi-story parking garage adjacent to residential buildings to mitigate the impact of noise on nearby residences. These solutions block and absorb sound, reducing noise transmission to the surrounding environment.

460. In constructing a high-performance home, which advanced framing technique is pivotal to increase insulation levels and reduce thermal bridging?
a. Utilizing double stud walls with a wider spacing
b. Employing traditional 16-inch on-center framing
c. Reducing the amount of header insulation
d. Avoiding the use of continuous insulation

Answer: a. Utilizing double stud walls with a wider spacing
Explanation: Utilizing double stud walls with wider spacing is crucial in constructing high-performance homes as it allows for increased insulation levels, reducing thermal bridging and enhancing the thermal efficiency of the building envelope.

461. While framing a multi-story residential building, a contractor decides to employ optimal value engineering. Which of the following considerations is vital in this approach?
a. Maximizing the use of structural lumber
b. Aligning windows and doors with the framing members
c. Avoiding insulation in the cavity spaces
d. Reducing the spacing between framing members

Answer: b. Aligning windows and doors with the framing members
Explanation: In optimal value engineering, aligning windows and doors with framing members is essential as it optimizes the placement of framing members, reduces waste, and maximizes the space available for insulation, improving overall energy efficiency.

462. While implementing advanced framing in a large commercial project, what framing member alteration is vital to minimize thermal bridging and enhance energy efficiency?
a. Employing corner studs with minimal spacing
b. Increasing the number of jack studs
c. Incorporating insulated three-stud corners
d. Avoiding the use of insulating sheathing

Answer: c. Incorporating insulated three-stud corners
Explanation: Incorporating insulated three-stud corners is pivotal as it minimizes thermal bridging and allows for better insulation, contributing to the enhancement of energy efficiency in the building.

463. In a project aiming for significant energy savings, which technique is crucial for reducing lumber use and maximizing insulation within the walls?
a. Employing 16-inch on-center stud spacing
b. Using ladder blocking at inline framing intersections
c. Avoiding insulated headers
d. Maximizing the use of solid blocking

Answer: b. Using ladder blocking at inline framing intersections
Explanation: Using ladder blocking at inline framing intersections is crucial as it reduces the use of lumber and maximizes the insulation within the walls, contributing to energy savings and improved thermal performance of the building.

464. During a residential project utilizing advanced framing techniques, which method is essential for ensuring structural integrity while optimizing material use?
a. Employing single top plates with aligned framing members
b. Increasing the number of cripple studs
c. Avoiding the use of shear walls
d. Ignoring the placement of structural sheathing

Answer: a. Employing single top plates with aligned framing members
Explanation: Employing single top plates with aligned framing members is essential as it optimizes material use without compromising the structural integrity of the building, by ensuring a continuous load path.

465. In a high-rise construction project aimed at sustainability, which framing technique is crucial for reducing material costs and waste, while maintaining structural strength?
a. Reducing the number of king studs and utilizing multiple jack studs
b. Incorporating stacked framing and aligned openings
c. Avoiding the use of rim boards
d. Ignoring thermal bridging considerations

Answer: b. Incorporating stacked framing and aligned openings
Explanation: Incorporating stacked framing and aligned openings is crucial as it maintains structural strength, reduces material costs, and waste by optimizing the placement of framing members and openings in alignment with each other.

466. While framing a commercial building, which of the following advanced framing techniques is crucial to reduce heat loss through the wall assemblies?
a. Increasing the use of framing members
b. Incorporating insulated headers sized to the load
c. Employing single headers with no insulation
d. Ignoring the alignment of framing members

Answer: b. Incorporating insulated headers sized to the load
Explanation: Incorporating insulated headers that are correctly sized to the load is crucial as it helps in reducing heat loss through wall assemblies by optimizing the amount of structural material and allowing for more insulation.

467. While implementing advanced framing techniques in a residential project, which consideration is vital to enhance structural stability and reduce lumber usage?
a. Employing double top plates
b. Incorporating inline framing with structural sheathing
c. Avoiding the installation of floor joists
d. Maximizing the use of solid sawn lumber

Answer: b. Incorporating inline framing with structural sheathing
Explanation: Incorporating inline framing with structural sheathing is vital as it enhances structural stability by creating a continuous load path and helps in reducing lumber usage, optimizing material use, and improving energy efficiency.

468. In a scenario where a contractor is building a passive house, which advanced framing technique is essential to minimize thermal breaks and optimize insulation?
a. Employing closely spaced framing members
b. Ignoring the placement of insulation in cavity spaces
c. Utilizing insulated headers
d. Avoiding the use of advanced framing techniques

Answer: c. Utilizing insulated headers
Explanation: In constructing a passive house, utilizing insulated headers is essential as they minimize thermal breaks and allow for optimal insulation within the framing assembly, contributing to the achievement of high-energy efficiency standards.

469. In a project to construct an energy-efficient commercial building, what framing strategy is pivotal to reduce thermal bridging through the assembly?
a. Maximizing the use of structural lumber in the corners
b. Ignoring the alignment of doors and windows with framing members
c. Incorporating two-stud corners with insulation
d. Employing continuous solid blocking in the wall assemblies

Answer: c. Incorporating two-stud corners with insulation
Explanation: Incorporating two-stud corners with insulation is pivotal as it reduces thermal bridging through the assembly, allowing for better insulation and contributing to the overall energy efficiency of the commercial building.

470. While working on a project with sustainability goals, a contractor decides to implement a concrete solution that leverages recycled materials. Which innovative concrete type would best suit this requirement?
a. High-Performance Concrete
b. Autoclaved Aerated Concrete
c. Green Concrete
d. Reactive Powder Concrete

Answer: c. Green Concrete
Explanation: Green Concrete utilizes recycled waste materials as its components, aligning well with sustainability goals. It not only helps in reducing the carbon footprint but also efficiently utilizes waste, making it the most appropriate choice for such projects.

471. A contractor is tasked with constructing a building that requires enhanced durability and strength. Which innovative concrete solution should be employed to meet these needs?
a. Lightweight Concrete
b. Self-Consolidating Concrete
c. Reactive Powder Concrete
d. Pervious Concrete

Answer: c. Reactive Powder Concrete
Explanation: Reactive Powder Concrete (RPC) is known for its enhanced durability and superior strength. It employs micro steel fibers and has a very low water-to-cement ratio, making it an ideal solution for projects demanding high strength and durability.

472. In a region with high precipitation, a contractor decides to implement a concrete solution that allows water to pass through, reducing runoff. What type of concrete should be used?
a. Self-Compacting Concrete
b. Pervious Concrete
c. Polymer Concrete
d. High Density Concrete

Answer: b. Pervious Concrete
Explanation: Pervious concrete allows water to pass through its porous structure, reducing runoff and allowing water to reach the ground. This is particularly useful in areas with high precipitation to manage water effectively.

473. To speed up the construction process, a contractor chooses a concrete solution that does not require vibration for placing and compacting. Which innovative concrete solution would be the best fit?
a. Lightweight Concrete
b. Self-Consolidating Concrete
c. Shotcrete
d. Ferrocement

Answer: b. Self-Consolidating Concrete
Explanation: Self-Consolidating Concrete has high fluidity and can spread into place under its weight, making it an ideal choice for speeding up construction processes as it eliminates the need for mechanical vibration for placing and compacting.

474. A contractor is working on a project in a seismic zone and needs to implement a concrete solution that can endure high impacts and vibrations. Which concrete type would be optimal for this project?
a. Engineered Cementitious Composite
b. High Density Concrete
c. Reactive Powder Concrete
d. Pervious Concrete

Answer: a. Engineered Cementitious Composite
Explanation: Engineered Cementitious Composite (ECC) is known for its strain-hardening and multiple cracking characteristics, making it highly suitable for areas prone to seismic activity, vibrations, and high impacts due to its enhanced ductility and damage tolerance.

475. In a marine project, which innovative concrete solution is crucial for resisting the corrosive action of salts and to ensure longevity of the structure?
a. Polymer Concrete
b. Shotcrete
c. Lightweight Concrete
d. Reactive Powder Concrete

Answer: a. Polymer Concrete
Explanation: Polymer concrete, with its corrosion-resistant properties, is ideal for marine environments where structures are exposed to corrosive salts. Its resistance to chemical attack ensures the durability and longevity of marine structures.

476. A contractor is required to use a concrete solution that can be applied pneumatically to any type or shape of surface, which type of concrete is best suited for this?
a. Ferrocement
b. Self-Consolidating Concrete
c. Shotcrete
d. Pervious Concrete

Answer: c. Shotcrete
Explanation: Shotcrete can be applied pneumatically and is especially suited to adhere to any type or shape of surface, making it versatile and ideal for complex geometrical structures.

477. In a scenario where rapid construction and early strength gain are pivotal, which innovative concrete solution should a contractor opt for?
a. High-Performance Concrete
b. Rapid Set Concrete
c. Lightweight Concrete
d. Autoclaved Aerated Concrete

Answer: b. Rapid Set Concrete
Explanation: Rapid Set Concrete is designed for quick setting and early strength gain, making it ideal in scenarios where rapid construction is essential, allowing for reduced downtime and increased productivity.

478. For a project that aims to optimize thermal insulation and fire resistance, which type of innovative concrete should be implemented?
a. High Density Concrete
b. Autoclaved Aerated Concrete
c. Reactive Powder Concrete
d. Polymer Concrete

Answer: b. Autoclaved Aerated Concrete

Explanation: Autoclaved Aerated Concrete (AAC) has excellent thermal insulation and fire resistance properties due to its porous structure, making it optimal for projects with thermal and fire resistance requirements.

479. A contractor is working on a building renovation project with space constraints. Which innovative concrete solution is critical to reduce the weight of the structure without compromising its strength?
a. Lightweight Concrete
b. High-Performance Concrete
c. Reactive Powder Concrete
d. Engineered Cementitious Composite

Answer: a. Lightweight Concrete. Explanation: Lightweight Concrete, with its reduced density, is crucial in scenarios with space and load constraints as it reduces the weight of the structure while maintaining adequate strength, making it suitable for renovations and buildings with space limitations.

480. A contractor firm is looking to improve its project efficiency by using a project management tool that allows for real-time collaboration and tracking of project progress. Which of the following would be the best fit?
a. Microsoft Excel
b. Trello
c. Microsoft Project
d. Google Sheets

Answer: b. Trello. Explanation: Trello allows for real-time collaboration and tracking of project progress with a highly visual and intuitive interface, making it suitable for improving project efficiency.

481. A contracting company is facing a lawsuit due to a breach of contract. Which of the following steps should be taken first by the management to resolve the dispute amicably?
a. Filing a counter lawsuit
b. Seeking mediation
c. Proceeding to arbitration
d. Ignoring the lawsuit

Answer: b. Seeking mediation. Explanation: Mediation is often the first step in dispute resolution as it is a less adversarial and more collaborative process to understand and resolve the differences between the involved parties.

482. A contractor is trying to determine the optimal markup to ensure the profitability of a project. Which of the following factors is crucial to consider when deciding the markup?
a. Overhead Costs
b. Competitors' Bids
c. Project Duration
d. Subcontractor Quotes

Answer: a. Overhead Costs

Explanation: Overhead costs are critical to determining markup as they represent the ongoing business expenses not directly attributed to creating a product or service. Ensuring overhead costs are covered is crucial to maintaining profitability.

483. Which of the following is a crucial business management practice for contractors to ensure the project stays within budget and on schedule?
a. Project Cost Management
b. Resource Leveling
c. Scope Creep Management
d. Time Tracking

Answer: a. Project Cost Management. Explanation: Project Cost Management is crucial to control costs and ensure that the project stays within budget. It involves estimating, budgeting, and controlling costs so the project can be completed within the approved budget.

484. Which legal business structure is typically best suited for a small contracting business wanting to protect personal assets but avoid double taxation?
a. Sole Proprietorship
b. Limited Liability Company (LLC)
c. Corporation
d. Partnership

Answer: b. Limited Liability Company (LLC). Explanation: An LLC provides the personal asset protection of a corporation and avoids double taxation by allowing profits/losses to be passed through directly to the owners.

485. To maintain a consistent cash flow and meet financial obligations, which business management strategy is crucial for contractors to implement?
a. Efficient Billing Cycle
b. Detailed Contract Review
c. Rigorous Project Planning
d. Strategic Marketing

Answer: a. Efficient Billing Cycle. Explanation: An efficient billing cycle is crucial for maintaining consistent cash flow by ensuring that invoices are sent out and paid in a timely manner, thereby enabling the contractor to meet financial obligations.

486. In a scenario where a contractor has to manage multiple projects simultaneously, which of the following would be most beneficial to ensure optimal resource allocation and scheduling?
a. Resource Leveling
b. Fast Tracking
c. Time Blocking
d. Crashing

Answer: a. Resource Leveling

Explanation: Resource Leveling is a technique used to examine unbalanced use of resources (usually manpower) and adjust project activities to optimize the allocation of resources, avoiding overallocation and ensuring smooth project execution.

487. A contractor is considering implementing a Customer Relationship Management (CRM) system to improve client relationships and increase repeat business. Which of the following benefits is directly associated with the implementation of a CRM system?
a. Cost Reduction
b. Enhanced Communication
c. Improved Scheduling
d. Budget Optimization

Answer: b. Enhanced Communication

Explanation: A CRM system centralizes customer information and enhances communication by allowing contractors to easily access and manage customer information, thereby improving client relationships and potentially increasing repeat business.

488. A contractor aiming to minimize risks associated with subcontractors should prioritize which of the following strategies?
a. Hiring the least expensive subcontractor
b. Extensive subcontractor vetting
c. Minimizing subcontractor involvement
d. Seeking multiple bids for comparison

Answer: b. Extensive subcontractor vetting

Explanation: Extensive vetting of subcontractors, including checking references, experience, financial stability, and certifications, is crucial to minimize risks associated with subcontractor performance, reliability, and compliance.

489. When facing a complex project with high uncertainty, which of the following contract types is most suitable to protect the contractor's interests?
a. Fixed-Price Contract
b. Cost-Plus Contract
c. Unit Price Contract
d. Lump Sum Contract

Answer: b. Cost-Plus Contract

Explanation: In a Cost-Plus Contract, the contractor is reimbursed for allowable or defined costs and receives a fee, usually a percentage of the costs, minimizing the financial risk associated with unforeseen complications or increased costs in complex projects.

490. A contracting company is seeking to establish a business structure that allows for separate legal entities and the ability to easily transfer ownership. Which business structure would be most suitable?
a. Sole Proprietorship
b. General Partnership
c. Corporation
d. Limited Liability Partnership

Answer: c. Corporation
Explanation: Corporations are considered separate legal entities from their owners, protecting personal assets. They also allow for easy transfer of ownership through the sale of stock and provide options for raising capital.

491. In a situation where two contractors with distinct specialties are forming a business together and wish to share management responsibilities, which business structure might be suitable?
a. Sole Proprietorship
b. Limited Liability Company (LLC)
c. Limited Partnership
d. General Partnership

Answer: d. General Partnership
Explanation: A General Partnership allows all partners to share in the management responsibilities and profits of the business, and it may be suitable for contractors with distinct specialties collaborating together.

492. A contractor who wishes to retain full control of the business while enjoying liability protection would likely opt for which business structure?
a. Corporation
b. Sole Proprietorship
c. Limited Liability Company (LLC)
d. General Partnership

Answer: c. Limited Liability Company (LLC)
Explanation: An LLC offers liability protection to the owner, like a corporation, while allowing the owner to retain full control over the business operations, similar to a Sole Proprietorship.

493. Which business structure is characterized by easy establishment but offers no protection of personal assets?
a. Sole Proprietorship
b. Corporation
c. Limited Liability Company (LLC)
d. Limited Partnership

Answer: a. Sole Proprietorship
Explanation: Sole Proprietorship is easy to establish with minimal formalities but does not provide protection of the owner's personal assets against business liabilities or debts.

494. Which of the following business structures typically involves a double taxation scenario?
a. Limited Liability Company (LLC)
b. Sole Proprietorship
c. Corporation
d. General Partnership

Answer: c. Corporation
Explanation: Corporations typically face double taxation, where the company's profits are taxed at the corporate level, and any dividends distributed to shareholders are taxed again at the individual level.

495. A contractor looking for a flexible management structure and pass-through taxation should consider which business structure?
a. Corporation
b. Sole Proprietorship
c. Limited Liability Company (LLC)
d. General Partnership

Answer: c. Limited Liability Company (LLC)
Explanation: LLCs offer a flexible management structure and benefit from pass-through taxation, where profits and losses are reported on the owner's individual tax return, avoiding double taxation.

496. In a scenario where contractors want to pool resources for a large-scale project without forming a new legal entity, they might consider establishing a:
a. Joint Venture
b. Corporation
c. Sole Proprietorship
d. Limited Partnership

Answer: a. Joint Venture
Explanation: A Joint Venture allows multiple contractors to pool resources for a specific project or time frame without creating a new permanent legal entity, thus retaining their original business structures.

497. When contractors wish to establish a business structure with distinct roles, where one or more partners are not involved in the day-to-day management but contribute capital, they might opt for a:
a. Sole Proprietorship
b. General Partnership
c. Limited Partnership
d. Corporation

Answer: c. Limited Partnership
Explanation: In a Limited Partnership, there are General Partners who manage the business and Limited Partners who contribute capital but do not participate in management and have limited liability.

498. For a contractor who primarily values ease of setup and operation in a business structure, which option would likely be the most appealing?
a. Corporation
b. Limited Partnership
c. Sole Proprietorship
d. Limited Liability Company (LLC)

Answer: c. Sole Proprietorship
Explanation: Sole Proprietorship is the simplest business structure to set up and operate, with fewer formalities, paperwork, and expenses, making it appealing to those who value ease of setup and operation.

499. A contracting company engaged in high-risk projects is looking for a business structure that offers maximum protection for its owners' personal assets. The company should consider forming a:
a. Sole Proprietorship
b. General Partnership
c. Corporation
d. Limited Liability Partnership

Answer: c. Corporation
Explanation: A Corporation provides the most protection to its owners' personal assets as it is considered a separate legal entity, isolating business liabilities and debts from the owners.

500. A construction company is looking to optimize their supply chain to improve project timelines. Which approach will most likely yield the best results in terms of project delivery time?
a. Lean Management
b. Critical Path Method
c. Six Sigma
d. Just-In-Time Inventory

Answer: b. Critical Path Method
Explanation: Critical Path Method (CPM) is specifically designed to optimize project scheduling and reduce timelines, making it more suitable for improving project delivery time in construction operations.

501. When facing unexpected delays in construction due to resource shortages, which of the following management strategies is most effective in maintaining project schedules?
a. Fast Tracking
b. Resource Leveling
c. Crashing
d. Scope Reduction

Answer: a. Fast Tracking
Explanation: Fast Tracking involves performing critical activities in parallel to overcome delays, maintaining project schedules when unexpected resource shortages occur.

502. For a construction company seeking to improve process quality and reduce defects, which management approach would be most suitable?
a. Lean Management
b. Six Sigma
c. Critical Path Method
d. Just-In-Time Inventory

Answer: b. Six Sigma
Explanation: Six Sigma focuses on improving process quality, reducing variability and defects, making it suitable for companies aiming for quality improvement in their operations.

503. A contractor is faced with the challenge of managing multiple projects with shared resources. What strategy should be employed to allocate resources efficiently without compromising on project timelines?
a. Resource Smoothing
b. Resource Leveling
c. Fast Tracking
d. Crashing

Answer: b. Resource Leveling
Explanation: Resource Leveling is used to allocate shared resources optimally between multiple projects, preventing overallocation and ensuring efficient utilization without affecting timelines.

504. To minimize waste and reduce costs during a project, a construction company should implement:
a. Lean Management
b. Six Sigma
c. Critical Path Method
d. Crashing

Answer: a. Lean Management
Explanation: Lean Management focuses on reducing waste, optimizing workflow, and decreasing costs, making it suitable for construction companies aiming for cost reduction and waste minimization in their projects.

505. In a scenario where a project is behind schedule, and the contractor needs to expedite project completion, which technique should be applied?
a. Resource Leveling
b. Fast Tracking
c. Crashing
d. Scope Reduction

Answer: c. Crashing
Explanation: Crashing involves allocating more resources or increasing working hours to expedite project completion when behind schedule, even though it may increase costs.

506. To ensure continuous supply of materials without holding excess inventory, a construction company should implement:
a. Lean Management
b. Just-In-Time Inventory
c. Six Sigma
d. Resource Leveling

Answer: b. Just-In-Time Inventory
Explanation: Just-In-Time Inventory aims to minimize inventory holding costs by ensuring materials arrive exactly when needed, maintaining a continuous supply without excess inventory.

507. A construction manager is looking to identify all the possible risks that might affect the project and determine their impact. This process is known as:
a. Risk Identification
b. Risk Assessment
c. Risk Mitigation
d. Risk Monitoring

Answer: b. Risk Assessment
Explanation: Risk Assessment involves identifying all possible risks, analyzing their potential impact, and determining the likelihood of their occurrence to develop suitable management strategies.

508. For maximizing productivity and optimizing workflow in construction operations, implementing which of the following would be most beneficial?
a. Lean Management
b. Six Sigma
c. Resource Leveling
d. Critical Path Method

Answer: a. Lean Management
Explanation: Lean Management is designed to optimize workflow, reduce waste, and improve overall productivity by eliminating non-value-adding activities in construction operations.

509. In a high-risk construction project, the emphasis should be on:
a. Risk Avoidance
b. Risk Mitigation
c. Risk Transfer
d. Risk Acceptance

Answer: b. Risk Mitigation
Explanation: In high-risk construction projects, Risk Mitigation strategies should be developed to reduce the likelihood and impact of identified risks, ensuring project success and safety.

510. A subcontractor defaults during a critical phase of project completion. Which clause allows a general contractor to directly employ an alternative entity to complete the subcontractor's work?
a. Pay-if-Paid Clause
b. Liquidated Damages Clause
c. Right-to-Cure Clause
d. No Damage for Delay Clause

Answer: c. Right-to-Cure Clause
Explanation: The Right-to-Cure Clause allows a general contractor to remedy the default by employing alternative means or entities to complete the work without terminating the subcontract.

511. In the event of disputes related to project delays, which contractual provision serves to prevent subcontractors from claiming additional costs or time extensions?
a. No Damage for Delay Clause
b. Liquidated Damages Clause
c. Pay-if-Paid Clause
d. Right-to-Cure Clause

Answer: a. No Damage for Delay Clause
Explanation: This clause limits subcontractors from claiming additional costs or time extensions due to delays, typically allowing only an extension of time to complete the project.

512. A construction contract states that any additional work requested by the owner not covered under the original scope will be billed separately. This refers to:
a. Quantum Meruit
b. Express Contract
c. Implied Contract
d. Unilateral Contract

Answer: a. Quantum Meruit
Explanation: Quantum Meruit allows contractors to receive compensation for services provided that were not included under the original contractual scope, preventing unjust enrichment.

513. In a Design-Bid-Build project delivery method, who bears the majority of the design risk?
a. Owner
b. Contractor
c. Designer
d. Subcontractor

Answer: c. Designer
Explanation: In Design-Bid-Build, the designer or architect is primarily responsible for the design, bearing the majority of the design risk related to errors, omissions, or non-compliance with applicable laws and regulations.

514. Which legal doctrine asserts that a contractor performing work in a visible and open manner without the owner's objection is presumed to have the owner's approval?
a. Quantum Meruit
b. Ostensible Agency
c. Mechanic's Lien
d. Laches

Answer: b. Ostensible Agency
Explanation: Ostensible Agency implies that the contractor, visibly and openly performing work without objection from the owner, is presumed to be acting with the owner's approval or authority.

515. In a construction project, a contractor deviates from the provided plans without approval, leading to defective work. This is a breach of:
a. Warranty of Merchantability
b. Implied Warranty
c. Warranty of Fitness
d. Express Warranty

Answer: b. Implied Warranty. Explanation: Implied Warranty in construction refers to the understanding that the work will be performed according to the plans and specifications; deviation without approval leads to a breach of this warranty.

516. A contractor is tasked with constructing a building on a soil type not previously disclosed that requires additional work and costs. The contractor can claim:
a. Differing Site Conditions
b. Force Majeure
c. Implied Warranty
d. Express Contract

Answer: a. Differing Site Conditions. Explanation: Differing Site Conditions clause allows contractors to claim additional costs and time due to unforeseen or undisclosed site conditions affecting the construction work.

517. To avoid claims and disputes related to delay damages, contracts often include clauses defining acceptable delays as:
a. Concurrent Delays
b. Critical Delays
c. Excusable Delays
d. Compensable Delays

Answer: c. Excusable Delays

Explanation: Excusable Delays are specified in contracts to define circumstances beyond the control of the contracted parties, which justify an extension of the project completion time without additional compensation.

518. In the absence of a written agreement, which type of contract is inferred from the conduct of the parties involved in a construction project?
a. Express Contract
b. Unilateral Contract
c. Implied Contract
d. Bilateral Contract

Answer: c. Implied Contract

Explanation: An Implied Contract is inferred from the actions or conduct of the parties involved, in the absence of a written or verbal agreement, creating mutual obligations.

519. If an owner intentionally interferes with a contractor's work, causing delays or additional costs, this is considered:
a. Constructive Change
b. Implied Warranty Breach
c. Force Majeure
d. Tortious Interference

Answer: d. Tortious Interference. Explanation: Tortious Interference occurs when a third party, such as an owner, intentionally disrupts or interferes with a contractor's performance, leading to delays or additional costs. The contractor may seek damages for such interference.

520. When a contractor agrees to complete a specific project by a certain date and receives a specified sum upon completion, this arrangement is considered a:
a. Cost-Plus Contract
b. Time and Material Contract
c. Lump Sum Contract
d. Unit Price Contract

Answer: c. Lump Sum Contract Explanation: A Lump Sum Contract involves an agreement where the contractor is to complete work for a stated sum of money, inclusive of all his costs, profits, and direct and indirect expenses.

521. During a project, a contractor decides to use a different construction method than initially agreed upon in the contract, asserting it will yield the same result. The contractor is claiming:
a. Equivalence
b. Substitution
c. Modification
d. Alteration

Answer: a. Equivalence
Explanation: Claiming equivalence implies the contractor is proposing a different method or material, asserting it provides equal performance and quality to what was specified in the contract.

522. If a party to a contract has failed to fulfill an obligation and the other party is relieved from their corresponding performance, this scenario illustrates:
a. Rescission
b. Discharge
c. Excuse of Performance
d. Assignment

Answer: c. Excuse of Performance
Explanation: Excuse of Performance occurs when one party's failure to fulfill their contractual obligation relieves the other party from their corresponding contractual duties.

523. A contractor accepts a project knowing that specific local building codes must be followed but realizes midway that adherence will cause substantial loss. The contractor is experiencing:
a. Impossibility of Performance
b. Frustration of Purpose
c. Mutual Mistake
d. Impracticability

Answer: d. Impracticability
Explanation: Impracticability arises when unforeseen circumstances make the contractual obligations substantially more burdensome, expensive, or difficult than initially anticipated.

524. The owner and contractor enter into a contract with agreed prices for units of work but without a committed total price. This is called a:
a. Cost-Plus Contract
b. Unit Price Contract
c. Lump Sum Contract
d. Guaranteed Maximum Price Contract

Answer: b. Unit Price Contract
Explanation: In a Unit Price Contract, the contractor is paid a specified amount per unit for each part of the work, and the total price is not fixed at the time of contracting.

525. A subcontractor on a project is making a claim due to non-payment. They allege that the general contractor's inability to manage their contractual relationships is to blame. This can be considered:
a. Breach of Warranty
b. Breach of Contract
c. Breach of Good Faith and Fair Dealing
d. Breach of Express Condition

Answer: c. Breach of Good Faith and Fair Dealing
Explanation: Every contract implies a covenant of good faith and fair dealing, requiring parties to act in a way that does not destroy the other party's rights to receive benefits under the contract.

526. When a contract includes terms allowing the owner to change contractual components without affecting other terms, this is known as a:
a. Flexible Provision
b. Modifying Clause
c. Change Order
d. Variable Term

Answer: c. Change Order
Explanation: A Change Order is a contractual provision allowing one party, typically the owner, to make changes to the agreed-upon work without altering the rest of the contract terms.

527. To prevent potential disputes regarding unknown site conditions, contracts generally include a clause that outlines the procedures and allowances for dealing with them. This clause is known as:
a. Site Investigation Clause
b. Differing Site Conditions Clause
c. Subsurface Conditions Clause
d. Geological Variance Clause

Answer: b. Differing Site Conditions Clause
Explanation: This clause outlines the protocol for addressing unexpected site conditions, allocating the risk and providing a mechanism for compensation and time extensions.

528. A contractor unknowingly enters into a contract with misrepresented facts. The contract can potentially be voided due to:
a. Duress
b. Mistake
c. Misrepresentation
d. Unconscionability

Answer: c. Misrepresentation
Explanation: When a party enters into a contract based on misrepresented facts, it can potentially void the contract as it lacks free and true consent.

529. If a contractor substantially performs their contractual obligations but with minor deviations, the contractor is entitled to:
a. Partial Payment
b. Full Payment minus Damages
c. Full Payment
d. No Payment

Answer: b. Full Payment minus Damages. Explanation: Substantial performance entitles the contractor to the contract price minus the amount needed to correct the deviations or deficiencies.

530. While reviewing the payroll reports, a contractor finds that an employee, who has been working overtime, has not received the overtime pay rate. Under the Fair Labor Standards Act (FLSA), overtime pay must be at least:
a. Same as the standard rate
b. Time and a half the standard rate
c. Double the standard rate
d. Triple the standard rate

Answer: b. Time and a half the standard rate. Explanation: The FLSA requires employers to pay non-exempt employees at least one and a half times their regular rate of pay when they exceed 40 hours of work in a workweek.

531. A contractor frequently hires a specific subcontractor based on the subcontractor's assurance that all its employees are eligible to work in the U.S. However, upon a surprise check, it's discovered that several of the subcontractor's workers are undocumented. Who could potentially be held liable for this violation?
a. Only the subcontractor
b. Only the contractor
c. Both the contractor and the subcontractor
d. Neither the contractor nor the subcontractor

Answer: c. Both the contractor and the subcontractor. Explanation: Under the Immigration Reform and Control Act (IRCA), both contractors and subcontractors can be held responsible for knowingly employing unauthorized workers.

532. At a construction site, a worker complains about potential hazards. Two weeks later, the worker is demoted. This can be seen as a violation of:
a. Whistleblower protection
b. Equal Employment Opportunity
c. Wage and hour laws
d. Family and Medical Leave Act

Answer: a. Whistleblower protection. Explanation: Whistleblower protection laws protect employees who disclose violations or unsafe conditions from retaliation by their employers.

533. Contractor A hires Contractor B for a project. Contractor B does not provide safety equipment to their employees, leading to an injury. According to the Occupational Safety and Health Act (OSHA), who is responsible for the safe conditions of employees?
a. Contractor A only
b. Contractor B only
c. Both Contractor A and B
d. The injured employee

Answer: c. Both Contractor A and B
Explanation: Under OSHA, both the hiring contractor and the subcontractor share responsibility for the safety of workers on a job site.

534. Jane, a supervisor in a construction firm, frequently makes unwelcome advances towards a male subordinate. This is an example of:
a. Hostile work environment
b. Wage discrimination
c. Wrongful termination
d. Equal Employment Opportunity violation

Answer: a. Hostile work environment
Explanation: Unwelcome advances, when severe or pervasive, can lead to a hostile work environment, a form of sexual harassment.

535. During a project, a contractor realizes that one of their subcontractors does not offer equal pay to male and female workers performing the same job. This is a potential violation of:
a. The Equal Pay Act
b. Title VII of the Civil Rights Act
c. The Age Discrimination in Employment Act
d. The Americans with Disabilities Act

Answer: a. The Equal Pay Act
Explanation: The Equal Pay Act prohibits wage discrimination based solely on gender when employees perform substantially equal work.

536. A construction firm wants to hire only employees aged 25-40, believing they will be more physically capable. This practice can be considered a violation of:
a. Fair Labor Standards Act
b. Equal Employment Opportunity Commission regulations
c. The Age Discrimination in Employment Act
d. Occupational Safety and Health Act

Answer: c. The Age Discrimination in Employment Act
Explanation: The Age Discrimination in Employment Act prohibits employment discrimination against individuals aged 40 and over.

537. A subcontractor on a project refuses to hire a qualified candidate solely due to the candidate's religious beliefs. This subcontractor is potentially violating:
a. The Americans with Disabilities Act
b. Title VII of the Civil Rights Act
c. The Age Discrimination in Employment Act
d. The Equal Pay Act

Answer: b. Title VII of the Civil Rights Act
Explanation: Title VII prohibits employment discrimination based on race, color, religion, sex, or national origin.

538. Carlos, an employee at a construction firm, requires a day off to observe a religious holiday. The employer refuses without examining possible accommodations. This scenario may be a breach of:
a. Fair Labor Standards Act
b. The Equal Pay Act
c. Title VII of the Civil Rights Act
d. The Age Discrimination in Employment Act

Answer: c. Title VII of the Civil Rights Act
Explanation: Title VII requires employers to reasonably accommodate an employee's religious beliefs or practices, unless doing so would impose an undue hardship.

539. An employee is diagnosed with a temporary disability due to an accident. He can perform the essential functions of his job with some modifications. If the employer refuses to make these accommodations, it may violate:
a. The Americans with Disabilities Act
b. Title VII of the Civil Rights Act
c. The Equal Pay Act
d. The Age Discrimination in Employment Act

Answer: a. The Americans with Disabilities Act
Explanation: The ADA requires employers to provide reasonable accommodations for employees with disabilities, unless it causes undue hardship.

540. A contracting company is considering purchasing a new piece of equipment. The contractor's accountant suggests using a depreciation method that will allocate the cost of the equipment equally over its useful life. This method is known as:
a. Double declining balance
b. Sum-of-the-years' digits
c. Units of production
d. Straight-line depreciation

Answer: d. Straight-line depreciation. Explanation: Straight-line depreciation spreads the cost of the asset equally over its useful life. This results in a consistent annual depreciation expense.

541. John's Construction has received a partial payment for a job, which is meant to cover the initial expenses. This type of payment is typically referred to as:
a. Retainage
b. A milestone payment
c. A down payment
d. An overhead payment

Answer: c. A down payment
Explanation: A down payment is an upfront payment made before the work begins or during the early stages, typically intended to cover initial costs.

542. BlueSky Contractors have total liabilities of $400,000 and owner's equity of $600,000. Given this, the company's total assets would be:
a. $200,000
b. $1,000,000
c. $1,600,000
d. $2,000,000

Answer: b. $1,000,000
Explanation: Using the accounting equation, Assets = Liabilities + Owner's Equity. Therefore, Assets = $400,000 + $600,000 = $1,000,000.

543. While reviewing the financial statements of a subcontractor, a contractor notices that the subcontractor's current ratio is below 1. This indicates:
a. The subcontractor is likely in a good short-term financial position.
b. The subcontractor might have difficulty paying off its short-term liabilities.
c. The subcontractor has more long-term assets than short-term assets.
d. The subcontractor is taking on more debt than equity.

Answer: b. The subcontractor might have difficulty paying off its short-term liabilities.
Explanation: A current ratio (current assets/current liabilities) below 1 suggests that the entity might not have enough assets to cover its short-term obligations.

544. Mason Contractors has a project that will cost $500,000 and is expected to generate returns of $600,000. What is the Net Present Value (NPV) of this project?
a. $100,000
b. $1,100,000
c. $500,000
d. -$100,000

Answer: a. $100,000
Explanation: NPV is calculated as the difference between the present value of cash inflows and the present value of cash outflows. In this case, $600,000 - $500,000 = $100,000.

545. A contractor is trying to determine the profitability of a project after all expenses, including taxes and interest. The contractor should look at the:
a. Gross profit margin
b. Operating margin
c. EBITDA margin
d. Net profit margin

Answer: d. Net profit margin
Explanation: The net profit margin represents the percentage of profit a company has earned from its total revenue after all expenses, including interest and taxes, have been deducted.

546. To secure a large contract, ABC Contractors is considering offering extended payment terms to a client. While this might improve sales, a potential drawback could be:
a. A reduction in tax liability.
b. A decrease in net profit margin.
c. An increase in accounts receivable days.
d. A decrease in depreciation expenses.

Answer: c. An increase in accounts receivable days.
Explanation: Extended payment terms might lead to delays in receiving payments, thereby increasing the average number of days that receivables remain outstanding.

547. During a busy construction season, XYZ Builders borrowed short-term loans to finance some of its operations. These types of loans are typically referred to as:
a. Mortgage loans
b. Bridge loans
c. Secured loans
d. Term loans

Answer: b. Bridge loans
Explanation: Bridge loans are short-term loans designed to provide temporary financing until a more permanent form of financing can be obtained.

548. StoneWall Inc. completed a project and the client held back 5% of the total payment until a later specified date to ensure all work was done satisfactorily. This is known as:
a. An advance payment
b. A milestone payment
c. Retainage
d. A performance bond

Answer: c. Retainage
Explanation: Retainage is a portion of the agreed-upon contract price deliberately withheld until the work is substantially complete to ensure the contractor completes the final details of the contract.

549. A contractor is determining the cost-plus-fixed-fee for a project. The estimated cost is $500,000 and the agreed-upon fee is 10%. The total price for the project would be:
a. $50,000
b. $450,000
c. $500,000
d. $550,000

Answer: d. $550,000
Explanation: The total price in a cost-plus-fixed-fee agreement is the cost plus the fee. In this case, the fee is 10% of $500,000, which is $50,000. Adding that to the estimated cost gives $550,000.

550. You're reviewing the financial statements of a subcontractor and notice a significant year-over-year increase in the account "Work in Progress (WIP)." This could indicate:
a. The subcontractor has been completing jobs ahead of schedule.
b. The subcontractor has taken on multiple large projects that aren't yet finished.
c. The subcontractor's revenue has decreased over the past year.
d. The subcontractor has been consistently underbidding on projects.

Answer: b. The subcontractor has taken on multiple large projects that aren't yet finished.
Explanation: A significant increase in WIP suggests the subcontractor has many ongoing projects, which have not yet reached a point where they can be billed in full.

551. After completing a construction project, LMN Builders reported a Deferred Tax Liability on their balance sheet. This typically results from:
a. Taxable income being higher than accounting income.
b. Accounting income being higher than taxable income.
c. Short-term debt financing.
d. A decrease in net operating losses.

Answer: b. Accounting income being higher than taxable income.
Explanation: When accounting income (book income) exceeds taxable income, it often results in a deferred tax liability. This means the company will owe more taxes in the future due to temporary differences between accounting and tax rules.

552. A contractor is determining the overhead rate for job costing. If the company has total overhead costs of $300,000 and direct labor costs of $1,000,000, the overhead rate would be:
a. 30%
b. 0.3%
c. 3%
d. 300%

Answer: a. 30%
Explanation: Overhead rate = (Total overhead costs / Direct labor costs) x 100 = ($300,000 / $1,000,000) x 100 = 30%.

553. When analyzing the year-end financial statements, you notice that ABC Contractors has a high Accounts Payable Turnover ratio. This could indicate:
a. ABC Contractors delays paying its suppliers.
b. ABC Contractors pays its suppliers promptly.
c. The company has a large amount of long-term debt.
d. The company's revenue has decreased over the past year.

Answer: b. ABC Contractors pays its suppliers promptly.
Explanation: A high Accounts Payable Turnover ratio suggests that a company pays off its suppliers at a faster rate.

554. XYZ Construction took a construction bond which ensures that they will complete a project according to the terms and conditions of a contract. This bond is known as:
a. Bid bond
b. Payment bond
c. Performance bond
d. Maintenance bond

Answer: c. Performance bond
Explanation: A performance bond guarantees that the contractor will perform the work as per the contract's terms and conditions.

555. A contractor reviews a job's ledger and realizes the costs attributed to the project have exceeded the budgeted amounts, but the project is only halfway complete. To gauge profitability, he should consider utilizing:
a. The percentage-of-completion method.
b. The completed-contract method.
c. Cash accounting.
d. Accrual accounting.

Answer: a. The percentage-of-completion method.
Explanation: The percentage-of-completion method recognizes revenues and expenses of long-term contracts based on the progress of the work, and it would be suitable to gauge profitability in this scenario.

556. The financial controller of Stone Builders realized they had underpaid their estimated quarterly taxes. This might result in:
a. A tax refund.
b. Understated liabilities.
c. Penalties or interest from the tax authorities.
d. An increase in working capital.

Answer: c. Penalties or interest from the tax authorities. Explanation: Underpaying estimated quarterly taxes can result in penalties or interest charges from tax authorities.

557. A contracting company decided to change its method of inventory valuation from FIFO (First-In, First-Out) to LIFO (Last-In, First-Out) during a period of rising prices. This change would:
a. Increase the company's net income.
b. Decrease the company's net income.
c. Have no effect on the company's net income.
d. Increase the company's tax liability.

Answer: b. Decrease the company's net income.
Explanation: During a period of rising prices, using LIFO will result in higher cost of goods sold and consequently lower net income compared to FIFO.

558. Brick & Mortar Contractors received an advance from a client for a project not yet started. In the company's balance sheet, this will be recorded as:
a. Accounts Receivable.
b. Work in Progress.
c. Deferred Revenue.
d. Accounts Payable.

Answer: c. Deferred Revenue.
Explanation: An advance received for a project not yet started is a liability for the company until the service is performed, hence it's recorded as Deferred Revenue.

559. At the end of the fiscal year, a contractor reviews financials and notes a substantial difference between taxable income and financial accounting income due to timing differences. This results in:
a. Permanent tax difference.
b. Deferred tax asset.
c. Deferred tax liability.
d. Both b and c.

Answer: d. Both b and c.
Explanation: Timing differences between taxable income and financial accounting income can result in either a deferred tax asset or liability, depending on the nature of the differences.

560. After a major storm hit the area, a contractor was called in to assess damage to a local school. Upon visiting, he noticed that while the roof was insured against such events, the surrounding fence which was also damaged was not. This exemplifies which risk management technique?
a. Risk retention
b. Risk transfer
c. Risk avoidance
d. Non-insurance transfer

Answer: a. Risk retention. Explanation: By not insuring the fence, the school implicitly retained the risk associated with its potential damage.

561. A contractor is working on a high-rise building. He's concerned about potential injuries from falling objects despite taking all precautionary measures. He decides to get insurance to cover potential lawsuits. This action is a form of:
a. Risk reduction
b. Risk transfer
c. Risk avoidance
d. Risk acceptance

Answer: b. Risk transfer
Explanation: By getting insurance, the contractor is transferring the financial consequences of the potential risk to an insurance company.

562. BlueSky Constructors have a policy that they will not work on any project located near a fault line due to the risk of earthquakes. This decision is an example of:
a. Risk retention
b. Risk transfer
c. Risk avoidance
d. Risk diversification

Answer: c. Risk avoidance. Explanation: BlueSky Constructors is avoiding taking on any risk associated with earthquake-prone areas by choosing not to work on projects near fault lines.

563. Mason Contractors recently faced a large financial loss due to theft at one of their sites. They decided to implement security systems at all their sites to minimize this risk in the future. This approach is known as:
a. Risk acceptance
b. Risk reduction
c. Risk avoidance
d. Risk transfer

Answer: b. Risk reduction. Explanation: By implementing security systems, Mason Contractors is reducing the likelihood or impact of future thefts.

564. A subcontractor working on a site suffered an injury due to the negligence of the general contractor. The subcontractor's Worker's Compensation Insurance covered the medical bills. Which principle of insurance is depicted here?
a. Indemnity
b. Insurable interest
c. Contribution
d. Subrogation

Answer: d. Subrogation
Explanation: Subrogation means that once the insurance company pays out a claim to the insured, they can seek compensation or recovery from the negligent party (in this case, the general contractor).

565. After a fire incident at a construction site, ABC Builders received a payout from their insurance company. However, they noticed that the payout was less than the total value of their loss. This could be due to:
a. Co-insurance clause
b. Beneficiary clause
c. An arbitration clause
d. Incontestability clause

Answer: a. Co-insurance clause
Explanation: A co-insurance clause may require the insured to bear a portion of the losses, resulting in the insurance payout being less than the total loss.

566. BrickLane Constructors has various insurance policies to protect against different risks. To ensure they aren't overpaying or receiving multiple compensations for a single claim, which principle should they be aware of?
a. Insurable interest
b. Subrogation
c. Indemnity
d. Contribution

Answer: d. Contribution
Explanation: The principle of contribution ensures that the insured doesn't profit from insurance by getting compensated from multiple insurers for the same claim.

567. XYZ Builders had a significant loss when one of their cranes was damaged during a project. The insurance company covered the claim but then decided to increase XYZ's premium rates for future policies. This adjustment of premiums is based on:
a. Loss frequency
b. Loss severity
c. Both a and b
d. Neither a nor b

Answer: c. Both a and b
Explanation: Insurance companies adjust premiums based on both the frequency and severity of claims.

568. Given the inherent risks in the construction industry, many contractors opt for a Comprehensive General Liability (CGL) insurance. Which of the following is NOT typically covered under CGL?
a. Injuries to employees
b. Property damage caused by construction activities
c. Bodily injuries or damage caused by completed products
d. Medical payments for injuries occurring on the business premises

Answer: a. Injuries to employees
Explanation: Injuries to employees are typically covered under Worker's Compensation insurance, not CGL.

569. A contractor working in multiple states is evaluating insurance options. He needs a policy that provides consistent coverage across states, regardless of individual state laws. He should consider:
a. A monoline policy
b. An occurrence policy
c. A claims-made policy
d. A broad-form policy

Answer: d. A broad-form policy
Explanation: Broad-form policies are designed to provide a consistent level of coverage, even in different jurisdictions with varying laws.

570. GreenTree Constructors has been experiencing increased turnover rates among young employees. They decide to implement a mentorship program where experienced workers mentor newcomers. This strategy is aimed at:
a. Wage leveling
b. Succession planning
c. Talent retention
d. Recruitment strategy enhancement

Answer: c. Talent retention
Explanation: Mentorship programs are often designed to help newcomers integrate and grow in the company, thereby aiding in talent retention.

571. A construction manager, James, notices that one of his teams consistently misses deadlines. On talking to the team lead, he discovers the team feels overworked. Which HR tool might James use to reassess workload distribution?
a. Employee engagement survey
b. Job analysis
c. Compensation survey
d. Exit interview

Answer: b. Job analysis
Explanation: Job analysis involves examining the tasks and responsibilities of a job, which can help James understand if the team is indeed overburdened.

572. BrickWall Builders has been getting feedback about safety concerns on their sites. The HR department decides to invest in safety training programs for all employees. This decision directly supports:
a. Compensation strategy
b. Employee relations
c. Organizational culture
d. Employee development

Answer: d. Employee development
Explanation: Training programs fall under employee development as they enhance skills and ensure that workers are equipped to handle their roles effectively and safely.

573. To address the gender wage gap in their organization, BuildTech Inc. decided to conduct a thorough review of salaries across roles. They aim to ensure equal pay for equal work. This approach is called:
a. Wage analysis
b. Pay equity analysis
c. Compensation benchmarking
d. Job evaluation

Answer: b. Pay equity analysis
Explanation: Pay equity analysis is a process to ensure that men and women are paid fairly and equally for performing the same or similar work.

574. XYZ Contractors is a large company with multiple projects across the state. They are considering implementing a software system that tracks time, attendance, benefits, and training for their workers. This system is known as:
a. Project Management Software
b. HR Information System (HRIS)
c. Enterprise Resource Planning (ERP)
d. Customer Relationship Management (CRM)

Answer: b. HR Information System (HRIS)
Explanation: An HRIS is designed to manage people, policies, and procedures. It usually handles activities such as time & attendance, benefits administration, and training records.

575. StoneBridge Constructions decided to implement a policy where employees could work four 10-hour days instead of five 8-hour days. This type of work arrangement is called:
a. Job sharing
b. Telecommuting
c. Compressed workweek
d. Flextime

Answer: c. Compressed workweek
Explanation: A compressed workweek allows employees to work their total number of agreed hours over fewer days.

576. Sarah, a project manager at BuildRight, feels there are communication gaps between her teams. HR suggests setting up weekly team meetings and monthly townhalls. This HR initiative aims to improve:
a. Employee morale
b. Organizational communication
c. Work-life balance
d. Team diversity

Answer: b. Organizational communication. Explanation: Regular meetings and townhalls aim to enhance communication within and between teams and the larger organization.

577. A construction company, HighRise Corp., wants to ensure they are hiring employees who fit their company culture and values. During the interview process, they place a high emphasis on behavioral questions. This is an example of:
a. Skill-based interviewing
b. Structured interviewing
c. Cognitive ability testing
d. Behavioral interviewing

Answer: d. Behavioral interviewing
Explanation: Behavioral interviewing focuses on understanding a candidate's past behavior in specific situations to predict their future behavior.

578. After a series of accidents on-site, PillarBuilders is focusing on safety training. They have also set up a system where employees can report near-miss incidents without fear of punishment. This system aims to:
a. Penalize non-compliance
b. Ensure employee accountability
c. Foster a safety culture
d. Enhance employee benefits

Answer: c. Foster a safety culture
Explanation: Reporting near-miss incidents helps in recognizing potential hazards before they cause harm, promoting a proactive safety culture.

579. BricksNMore is planning a new project and is assessing the manpower needed. The HR team looks at the past projects' performance metrics, current team strengths, and the complexity of the upcoming project. This process is known as:
a. Workforce planning
b. Job rotation
c. Employee onboarding
d. Employee appraisal

Answer: a. Workforce planning
Explanation: Workforce planning involves analyzing and forecasting the talent that a company needs to meet its objectives.

580. Mike, an experienced electrician at ElectraCorp, noticed that his pay is lower than a newer electrician. When he brought it up with management, they stated the new hire had specialized certifications. The scenario depicts a possible case of:
a. Unjust enrichment
b. Wage discrimination
c. Pay equity based on qualifications
d. Seniority-based wage system

Answer: c. Pay equity based on qualifications
Explanation: Wage discrepancies can arise due to specialized qualifications, and in this case, the newer electrician's certifications might justify higher pay.

581. At BuildWell Constructions, employees who complete two years with the company are sent for advanced skill training. This is an example of:
a. Onboarding training
b. Cross-training
c. Upskilling
d. Orientation

Answer: c. Upskilling
Explanation: Upskilling refers to training workers to perform more advanced roles or tasks in their existing occupation.

582. MasonWorks has a policy to promote employees based on their duration of service rather than performance metrics. This type of system is referred to as:
a. Merit-based
b. Seniority-based
c. Performance-driven
d. Needs-based

Answer: b. Seniority-based
Explanation: When promotions are determined by the length of service, it's called a seniority-based system.

583. Jenna, a forewoman at a construction site, wants to hire a new crew member. She is specifically looking for someone who can start immediately and has experience in high-rise construction. The best recruitment strategy for her would be:
a. College recruitment drives
b. Employee referrals
c. Job fairs
d. Using a staffing agency specializing in construction

Answer: d. Using a staffing agency specializing in construction
Explanation: A staffing agency, especially one specialized in construction, can quickly provide experienced candidates tailored to specific requirements.

584. Pinnacle Builders ensures that every new hire goes through a one-week program where they learn about the company's safety standards, culture, and basic job functions. This program is a form of:
a. Workshop
b. Seminar
c. Orientation
d. Internship

Answer: c. Orientation
Explanation: Orientation programs are designed to introduce new employees to the company and their roles.

585. StoneCraft Inc. offers an additional 10% hourly rate for employees who work after regular hours. This additional pay is known as:
a. Bonus
b. Commission
c. Overtime pay
d. Differential pay

Answer: d. Differential pay
Explanation: Differential pay is additional compensation for employees who work outside of regular working hours.

586. A worker at HighTower Constructions filed a complaint that they were not compensated for their lunch breaks, even though they often worked through them. Under federal labor laws, the company:
a. Is not required to pay as lunch breaks are non-compensable
b. Should pay for the break if the employee was working
c. Should only compensate if the break is less than 20 minutes
d. Is exempt from any wage laws regarding breaks

Answer: b. Should pay for the break if the employee was working
Explanation: Under the Fair Labor Standards Act (FLSA), short breaks (usually 20 minutes or less) are compensable. If an employee works during their lunch break, they should be compensated.

587. To ensure a diverse workforce, BridgeBuild Corp. collaborates with organizations that assist veterans, differently-abled individuals, and other minority groups in finding jobs. This is a tactic to:
a. Reduce wage bills
b. Achieve affirmative action goals
c. Ensure seniority-based hiring
d. Limit the talent pool

Answer: b. Achieve affirmative action goals
Explanation: Collaborating with such organizations helps in ensuring diversity and can be a part of a company's affirmative action initiative.

588. Nina, HR at ArchBuilders, is developing a program where employees are rotated between various jobs to increase their understanding and skills. This is known as:
a. Job enrichment
b. Job enlargement
c. Job rotation
d. Job specification

Answer: c. Job rotation
Explanation: Job rotation involves moving employees between different tasks to promote experience and variety.

589. Foundations Inc. is facing a lawsuit from a former employee who claims he was not paid for his overtime hours. In court, the company will need to present _____ as evidence of their compliance with wage laws.
a. Employee performance reviews
b. Timesheets and work records
c. Job advertisements
d. Training schedules

Answer: b. Timesheets and work records
Explanation: Timesheets and work records would provide evidence of the hours an employee worked and the compensation they received.

590. At the Radiant Tower construction site, the project manager instructs the crew to install a waterproofing membrane beneath the building's foundation. The primary purpose of this is:
a. Aesthetic appeal.
b. To prevent termite infestations.
c. To minimize the impact of soil erosion.
d. To prevent moisture penetration and protect the foundation.

Answer: d. To prevent moisture penetration and protect the foundation.
Explanation: Waterproofing membranes are primarily used beneath foundations to prevent moisture intrusion, which can compromise the foundation's structural integrity.

591. Jim, a site supervisor, notices that one of the poured concrete columns exhibits signs of honeycombing. The most probable cause for this is:
a. Rapid curing of the concrete.
b. Insufficient compaction during pouring.
c. Overuse of water in the mix.
d. Late removal of formwork.

Answer: b. Insufficient compaction during pouring.
Explanation: Honeycombing in concrete is often caused by inadequate compaction, which leads to voids and pockets in the finished concrete.

592. For a mixed-use skyscraper, the architect has chosen a "tube-in-tube" structural system. The main advantage of this system is:
a. Minimized construction time.
b. Enhanced resistance to lateral forces, such as wind.
c. Reduced use of construction materials.
d. Allows for larger open spaces within the building.

Answer: b. Enhanced resistance to lateral forces, such as wind.
Explanation: The "tube-in-tube" structural system is predominantly chosen for its enhanced resistance to lateral forces, crucial for skyscrapers.

593. ConstructCo is building on a site with a high water table. To ensure worker safety and project integrity, which method would they most likely use during excavation?
a. Install baffle boards.
b. Apply a load-bearing wall system.
c. Use of a dewatering system.
d. Incorporate air barriers.

Answer: c. Use of a dewatering system.
Explanation: On sites with high water tables, dewatering systems are employed to remove excess water from the excavation area.

594. While installing a metal roof, Clara recommends using a standing seam system. One of the key benefits of this system is:
a. Decreased roof pitch.
b. Increased aesthetic variability.
c. Enhanced waterproofing due to concealed fasteners.
d. Reduced weight of the roofing material.

Answer: c. Enhanced waterproofing due to concealed fasteners.
Explanation: Standing seam metal roofing systems have concealed fasteners, enhancing waterproofing by reducing potential entry points for water.

595. A contractor is hired to build a house in a seismic-prone area. To improve the structure's earthquake resistance, he should:
a. Add more masonry walls for stiffness.
b. Use inflexible joint connections.
c. Incorporate base isolators in the foundation.
d. Increase the weight of the roof.

Answer: c. Incorporate base isolators in the foundation.
Explanation: Base isolators act as shock absorbers and allow a building to move independently of ground shaking, significantly improving its earthquake resistance.

596. When constructing a multi-story car park, the builder decides to use precast concrete. This method offers the advantage of:
a. Longer curing times.
b. Enhanced on-site adaptability.
c. Speedier construction process.
d. Reduced need for finishes.

Answer: c. Speedier construction process.
Explanation: Precast concrete elements are manufactured off-site and brought to the construction site ready to be installed, greatly speeding up the construction process.

597. For a luxury hotel construction, the interior designer requests the use of a coffered ceiling. This ceiling type is:
a. A smooth, uninterrupted finish.
b. Divided into a grid of recessed squares or rectangles.
c. Predominantly used for soundproofing.
d. Made exclusively of metal.

Answer: b. Divided into a grid of recessed squares or rectangles.
Explanation: A coffered ceiling is characterized by its grid of recessed panels, offering both aesthetic and acoustic benefits.

598. A construction company is facing challenges with expansive soils at a building site. To manage this, the contractor could:
a. Reduce the depth of the foundation.
b. Use lighter construction materials.
c. Employ a floating foundation technique.
d. Avoid any ground treatment.

Answer: c. Employ a floating foundation technique.
Explanation: In areas with expansive soils, floating foundations can be used to allow the structure to "float" on the soil, minimizing potential damage.

599. When building a bridge in a coastal region, engineers decide to use weathering steel. This decision is mainly to:
a. Reduce the cost of the project.
b. Eliminate the need for painting.
c. Increase the weight-bearing capacity.
d. Make the structure flexible during storms.

Answer: b. Eliminate the need for painting.
Explanation: Weathering steel forms a protective rust patina, eliminating the need for regular painting and providing resistance against corrosion.

600. Tom, an electrician, is working in a commercial building. He needs to run a conduit for 480V power. Which type of conduit is most appropriate for this application in a potentially damp location?
a. PVC
b. EMT
c. RMC
d. Flex

Answer: c. RMC (Rigid Metal Conduit)
Explanation: RMC is a heavy-duty galvanized steel tubing used for electrical conductors. It has a thick wall and can be used in both exposed and concealed locations, including damp or wet areas.

601. In a new apartment complex, the plumber has to install a device to prevent backflow in a potable water distribution system. What is this device typically called?
a. Trap seal
b. P-trap
c. Vent stack
d. Backflow preventer

Answer: d. Backflow preventer
Explanation: A backflow preventer is designed to protect potable water supplies from contamination or pollution due to backflow.

602. Samantha, a contractor, is overseeing the installation of an exterior brick wall. She notices that the mason is placing vertical reinforcement every 6 feet. She knows this spacing is:
a. Too close and a waste of materials.
b. Standard for most brick walls.
c. Too wide and may cause structural issues.
d. Dependent on the height of the wall.

Answer: c. Too wide and may cause structural issues.
Explanation: For brick masonry, vertical reinforcement is typically spaced closer, often around 32 inches or depending on local building codes and wall height.

603. A roofer is working on a pitched roof and needs to calculate the total area to determine how much material to order. If each side of the roof measures 20 feet in width and 30 feet in length, what's the total area?
a. 600 sq. ft.
b. 1,200 sq. ft.
c. 2,400 sq. ft.
d. 900 sq. ft.

Answer: b. 1,200 sq. ft.
Explanation: Each side of the roof is 20 x 30 = 600 sq. ft. Since there are two sides, the total is 600 x 2 = 1,200 sq. ft.

604. In HVAC, when considering the installation of a split system, the component responsible for releasing the absorbed heat outside the building is known as:
a. The evaporator coil
b. The blower motor
c. The compressor
d. The condensing unit

Answer: d. The condensing unit
Explanation: In a split system, the condensing unit is the outdoor component that releases absorbed heat from the building to the outside environment.

605. During a residential renovation, a carpenter encounters a load-bearing wall. To create an open-concept design, he intends to remove it. What's essential for him to integrate before removal?
a. Temporary shoring
b. An adjacent wall for support
c. Moisture barriers
d. Thermal insulation

Answer: a. Temporary shoring
Explanation: Before removing a load-bearing wall, temporary shoring or bracing is required to support the load until a permanent beam or other structural element is in place.

606. A tile setter is laying tiles in a herringbone pattern. To ensure consistency in spacing between tiles, which tool is indispensable?
a. Tile nipper
b. Tile cutter
c. Grout float
d. Tile spacers

Answer: d. Tile spacers
Explanation: Tile spacers ensure uniform gaps between tiles which will later be filled with grout.

607. When installing a window in a coastal region, which feature is crucial for resisting strong wind forces like those from hurricanes?
a. Tempered glass
b. Impact-resistant glass
c. Frosted glass
d. Tinted glass

Answer: b. Impact-resistant glass
Explanation: In hurricane-prone areas, windows with impact-resistant glass are used to withstand flying debris and intense wind pressures.

608. A drywall installer needs to fix a large hole in a wall. Which of the following materials is best suited for this task?
a. Joint compound alone
b. Wall patch kit
c. Mesh tape and joint compound
d. Just drywall screws

Answer: c. Mesh tape and joint compound
Explanation: For larger holes, mesh tape provides a substrate for the joint compound to adhere to, ensuring a sturdy repair.

609. In a bathroom renovation, the contractor needs to waterproof the shower area. Which of the following is a common material used directly under tile installations for waterproofing?
a. Durock
b. RedGard
c. Sanded grout
d. Spackle

Answer: b. RedGard
Explanation: RedGard is a popular liquid waterproofing and crack prevention membrane used beneath tiles in wet areas.

610. While working on a commercial project, electrician Sam needs to connect metal parts or equipment to ensure electrical continuity. Which of the following best describes this process?
a. Isolation
b. Grounding
c. Insulation
d. Circuit breaking

Answer: b. Grounding
Explanation: Grounding ensures that any unintended electrical surges or faults will be directed into the earth, reducing the risk of electric shock or fire.

611. Plumber Jane is installing a new system in a high-rise apartment. To prevent water from siphoning back into the public water system, she should install:
a. A relief valve
b. A P-trap
c. A backflow preventer
d. A vent stack

Answer: c. A backflow preventer
Explanation: Backflow preventers are designed to stop water from flowing in the opposite direction and contaminating the public water system.

612. In an HVAC system, the component responsible for absorbing heat from inside a building and transferring it outside is:
a. The blower
b. The evaporator coil
c. The compressor
d. The condenser coil

Answer: b. The evaporator coil
Explanation: The evaporator coil absorbs heat from inside air, while the condenser coil releases that heat outside.

613. An electrician needs to calculate the load for a new circuit that will have six 150W bulbs. If the voltage is 120V, what will be the total current (in amps) of the circuit?
a. 7.5 A
b. 5 A
c. 3 A
d. 6.25 A

Answer: a. 7.5 A
Explanation: Total wattage = 6 x 150W = 900W. Using the formula P=IV (Power = Current x Voltage), Current (I) = P/V = 900W/120V = 7.5 A.

614. For an HVAC technician, when considering optimal efficiency and system longevity, the refrigerant charge should be:
a. As high as the system will allow.
b. Equal to the system's maximum capacity.
c. As recommended by the equipment manufacturer.
d. The same for all systems for consistency.

Answer: c. As recommended by the equipment manufacturer.
Explanation: Proper refrigerant charge as per the manufacturer's recommendation ensures the HVAC system operates efficiently and lasts longer.

615. A plumber is working in an older home and encounters piping made from a material that is grayish, lightweight, and was commonly used from the 1970s-1990s. This material is most likely:
a. PVC
b. Copper
c. Polybutylene
d. Galvanized steel

Answer: c. Polybutylene
Explanation: Polybutylene was a popular material for plumbing pipes from the 1970s through the 1990s, but it's now considered outdated due to potential failures.

616. In a commercial building, the electrician is asked to install emergency lights that activate when the primary power source is lost. Which device helps in achieving this?
a. Transformer
b. Inverter
c. Circuit breaker
d. Transfer switch

Answer: d. Transfer switch
Explanation: A transfer switch shifts the load to the emergency power source when the primary source fails, ensuring continuous power supply.

617. An HVAC technician is troubleshooting a malfunctioning AC system. He notices that the compressor is running, but the fan is not. This could be due to a faulty:
a. Thermostat
b. Evaporator coil
c. Capacitor
d. Expansion valve

Answer: c. Capacitor
Explanation: The capacitor provides the initial boost to start the motor and keeps it running. A faulty capacitor might prevent the fan motor from operating.

618. In a residential property, a plumber has to replace a section of pipe under the sink. For ease of installation and adjustment, they should use:
a. Rigid copper pipe
b. PEX tubing
c. ABS pipe
d. Compression fitting

Answer: d. Compression fitting
Explanation: Compression fittings allow for easy adjustments and can be used to connect different types of pipes without soldering or gluing.

619. An electrician is working on a circuit with three resistors connected in parallel. If one resistor fails open, what happens to the total resistance of the circuit?
a. It decreases
b. It remains the same
c. It increases
d. It becomes zero

Answer: c. It increases
Explanation: In parallel circuits, if one resistor fails or is removed, the total resistance increases because there's one less path for current to flow.

And there we have it. The journey through this study guide has been akin to weaving through the intricate tapestry of knowledge, where each thread represents a unique facet of the vast world of contracting. Remember, in the symphony of life, it's our dreams that compose the melody. There might be moments when you hit a false note, but that's where learning resides. Mistakes, after all, are just detours on the road to mastery.

Every hiccup, every uncertainty you've faced – or will face – is a stepping stone. Embrace them. Don't be disheartened by the shadows of doubt; they only mean there's a light shining somewhere nearby. And in times when the road gets foggy, lean on this guide. Let it be your compass, confirming your beliefs and dispelling those lingering suspicions.

As you approach your exam, remember the community you belong to. The countless professionals who, just like you, once stood at this threshold, their hearts brimming with aspirations, their minds buzzing with questions. And look at them now, constructing dreams, brick by brick, blueprint by blueprint.

Now, it's your turn. Picture yourself a few months, or even weeks from now - successfully navigating the professional landscape with confidence, your expertise recognized and respected. Bask in that pride, and let it fuel your motivation.

In the grand scheme of things, tests are just milestones. But the journey? Ah, that's where the real magic lies. Dive into the challenges, revel in the learning, and remember to support and uplift those who walk beside you.

Here's to the dreams you're chasing and the heights you're yet to reach. Best of luck.

Made in the USA
Las Vegas, NV
05 April 2024

88245780R00118